# Introduction to
# SOLID-STATE
# TELEVISION SYSTEMS

## Color and Black & White

PRENTICE-HALL SERIES IN ELECTRONIC TECHNOLOGY

Dr. Irving L. Kosow, editor

Charles M. Thomson, Joseph J. Gershon, and Joseph A. Labok, consulting editors

PRENTICE-HALL, INTERNATIONAL, INC., *London*
PRENTICE-HALL OF AUSTRALIA, PTY. LTD., *Sydney*
PRENTICE-HALL OF CANADA LTD., *Toronto*
PRENTICE-HALL OF INDIA PRIVATE LTD., *New Delhi*
PRENTICE-HALL OF JAPAN, INC., *Tokyo*

# Introduction to
# SOLID-STATE
# TELEVISION SYSTEMS

## Color and Black & White

**GERALD L. HANSEN**

*Grindle Enterprises, Inc.*
*San Diego, California*

**Prentice-Hall, Inc., Englewood Cliffs, New Jersey**

# PREFACE

The pace of modern technology and the dramatic miniaturization of electronic components within the past few years has produced a continuing revolution in the television industry. Television cameras and associated equipment have become amazingly compact in comparison to their predecessors. At the same time, power requirements have been reduced considerably and performance has been greatly improved. The widespread popularity of television systems in an ever-increasing number of applications attests to their reliability and versatility.

This book is written to provide a good working knowledge of the theory and operation of television systems. The basic theory is presented in a practical manner and the text is not bogged down in heavy mathematics or difficult concepts. The many examples of circuitry are typical of those found in modern solid-state television systems and should be easily grasped by the technically average individual who possesses a basic knowledge of transistor theory. In almost every case, component values and transistor types are shown in the circuit diagrams to eliminate vague references and to aid comprehension.

The book can be used in technical institutes, colleges, high schools, and industrial classes. It should be a valuable aid to those in industry who utilize or service television systems, and the broadcaster or cable television operator should also find it quite useful. Questions keyed to each chapter are included at the end of the book as an aid to testing comprehension of a class or individual.

The author is indebted to many companies and individuals for their contributions of material and assistance. Thanks go to Albion Optical Co., Ampex Corporation, Blonder-Tongue Laboratories, Inc., Cohu Electronics, Inc., CBS Laboratories, Conrac Corporation, Dynair Electronics, Inc.,

Electronic Industries Association, GPL Division of General Precision Systems, Inc., Motorola Inc., Radio Corporation of America, Rebikoff Underwater Products, Sylvania, Westinghouse, and Zoomar, Inc. Additional thanks to Dick Harmon of Cohu and Edward J. Dudley of RCA for their personal efforts and a wealth of material.

An especially valuable asset enjoyed by the author was his association with Mr. Gene Crow of Cohu Electronics, whose knowledge of the subject seems without limit.

GERALD L. HANSEN

*San Diego, California*

# TABLE OF CONTENTS

# Introduction to
# SOLID-STATE
# TELEVISION SYSTEMS

Color and Black & White

# TELEVISION
# APPLICATIONS

The usefulness of television and its impact upon human society is obvious. It has opened broad new avenues of approach in the fields of entertainment and news dissemination by its use in public broadcasting. Perhaps not so well known are the areas in science, industry, and education where the television camera has contributed immeasurably to man's versatility and his knowledge of his environment and himself.

The television camera is probably best described as an extension of the human eye. Film cameras have also been associated with this description, but television is more truly representative because of its ability to relay information instantaneously. Such "real time" image transmission is becoming more important as the complexity and pace of our society increases. Industry and science require the capability to view events occurring in extremely hazardous locations. The use of television systems in areas of atomic radiation, underwater environments, and the vacuum of space are but a mere sample of the many possibilities.

## BROADCASTING

Broadcasting is probably the most familiar use of television. Millions of television receivers in use around the world attest to its extreme popularity. Broadcast TV has been credited with changing the social patterns of a large segment of the human race. Certainly, many people are influenced by television in their social behavior, buying habits, political views, and the like.

The television cameras and associated equipment which are used in broadcast studios are usually quite sophisticated and expensive. The average broadcaster is not only concerned about the technical capabilities of his systems but, since he uses them as an art form, goes to great lengths to acquire a picture whose subtle qualities make it pleasing to view. For this reason television studios employ extensive lighting facilities and equipment whose special effects capabilities lend polish and glamor to the televised image.

The broadcaster in the United States must employ equipment which meets the standards of the FCC (Federal Communications Commission). The video signals that are generated and broadcast may not deviate from specific tolerances and must be monitored for conformance at all times. Such requirements, coupled to the broadcaster's inherent professionalism and commercial interests, result in the use of equipment that is considerably more elaborate than that used in most Closed Circuit Television (CCTV) applications.

The television cameras used in broadcast studios (Figure 1-1) are relatively large when compared to some of their closed circuit counterparts

COURTESY KOGO-TV, SAN DIEGO.

FIGURE 1-1. A monochrome broadcast camera.

(Figure 1-2). The greater physical size of the former is an outgrowth of the particular types of camera tubes used, the use of specialized circuitry not generally required in closed circuit applications, and the need for physical stability. The fact that larger cameras are also more impressive in appearance is not lost to the astute broadcaster who wishes his equipment to appear as "professional" as possible.

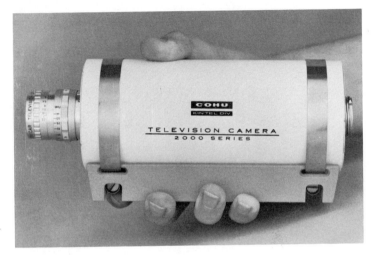

COURTESY COHU ELECTRONICS, INC.

FIGURE 1-2. A closed-circuit television camera head.

The advent of color television has created an additional dimension of realism for the broadcaster (Figure 1-3). While it is true that most people have become accustomed to the unnatural colorlessness of monochrome television, color television has become such an instant commercial success that black and white television transmissions for commercial purposes are becoming more rare with each passing day. The change in programming from monochrome to color has resulted in an increased complexity and cost for television stations that were unheard of in years past (Figure 1-4).

The wide variety of equipment which is used to support the television camera in a studio may take on many configurations. Additional items such as video tape recorders, film-chain cameras for display of movies and slides, lighting fixtures, audio systems, microwave transmitters and receivers, test equipment, and the like, crowd broadcast control rooms and studios. Items such as the video switcher (Figure 1-5), which is used to route the various video signals to different destinations, are generally quite complex and versatile and possess the capability to provide special effects and professional production techniques.

COURTESY KOGO-TV, SAN DIEGO.

FIGURE 1-3. Color television cameras add an additional dimension.

COURTESY KOGO-TV, SAN DIEGO.

FIGURE 1-4. Racks of equipment process and route the broadcast television signal.

While broadcast television is generally considered high in quality, it is quite severely limited in its ability to provide images of high resolution because of the restricted transmitting bandwidth which broadcasters have at their disposal. The bandwidth of the channel allocations in the United States is determined by the FCC and all broadcasters must conform to its requirements. Closed circuit television, on the other hand, is not broadcast and therefore may employ extremely wide bandwidths to obtain better resolution capabilities.

COURTESY KOGO-TV, SAN DIEGO.

FIGURE 1-5. Video switching equipment allows professional production techniques.

The popularity and success of commercial television is largely responsible for the existence of the greater number of closed circuit systems which are currently in use. The broadcast market provided the incentive and means for the development of equipment and techniques that would otherwise have required many years to develop (Figure 1-6).

COURTESY AMPEX CORPORATION.

FIGURE 1-6. A portable television camera and "back-pack" video tape recorder exemplify modern advances in equipment design.

## EDUCATION

Television will probably find its greatest challenges and achievements in the field of education. It is here that its unique ability to relate, inform, and teach find their ideal application. For example, in a classroom situation a system might provide the viewer with the combined instructional powers of motion pictures and slide transparencies, a microscope or telescope, and a teacher located many miles away, always giving the entire class a simultaneous, unobstructed view of the subject. Figure 1-7 illustrates a system which allows an entire assembly of students to participate in an experiment with their instructor and gain the immediacy of the results obtained.

Because broadcast television has proved beyond doubt the ability of television to educate, many of the techniques developed by broadcasters are now being applied in educational institutions. The continuing increase in the number of television systems manufactured, together with the introduction of transistors and integrated circuits, has served to provide systems that approach the quality of broadcast equipment at a continually lowering price. The use of such television systems is considered by many to be a definite turning point in educational procedures.

Consider the effect on a class of medical students able to watch the

COURTESY COHU ELECTRONICS, INC.

FIGURE 1-7. Television increases student understanding of experiments conducted before large assemblies.

movements of a surgeon's hands as he performs delicate surgery. It is also possible, by the use of flexible fiber optic devices, to enter various internal organs of the body and view directly their functions. Classic operations can be recorded on video tape and viewed again and again.

Another important area of education relying more and more upon television is teacher training. A major portion of a student teacher's time must be spent in actual classroom observation of the pupils. Cameras with remotely controlled pan-tilt units and zoom lenses may be unobtrusively mounted within the classroom and the scene viewed by a group of student teachers at a distant location. This use of television (Figure 1-8) enriches classroom observations by providing identical viewing situations for all without the disturbing effects of their presence within the classroom. Also, the instructor in charge of the student teachers may manipulate the cameras to emphasize those aspects of human behavior or classroom activity which he feels are important. When the student teacher conducts classes, his performance may be video taped for replay at a later time, allowing a unique opportunity for critical self-analysis.

While television can be an extremely effective device in educational applications, it can also be a poor investment if it is not properly utilized. It has generally been found that a television system that is simply purchased and handed to an unprepared educator, to use as he sees fit, will ultimately end up gathering dust in a storeroom. It is necessary to change teaching methods and habits and correlate the program to the device. In other words, teachers should be taught to use their new tool. Once this is done, there is ample evidence that the television camera lends amazing flexibility to the art of teaching.

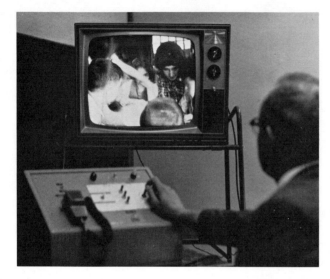

COURTESY COHU ELECTRONICS, INC.

FIGURE 1-8. Remote cameras give student teachers insight into pupil behavior.

## MEDICINE

Beyond its role in the education of medical students, television is also used extensively in many other areas of the medical field. It is a very effective means to monitor critically ill patients from a central location within a hospital. Television cameras also view patient reactions under radiation treatment that excludes the presence of a human observer (Figure 1-9).

The growing cost and complexity of medical equipment, coupled with the increasing trend toward specialization among doctors is responsible for the growth of huge medical complexes in many of the larger population centers of the United States. It is therefore necessary for patients from outlying, smaller communities to commute to these central locations for specialized diagnosis and treatment. Television systems with high-resolution capabilities may serve as a link between these two areas. Such systems allow the specialist to view X-rays and other pertinent data from the convenience of his office, and even converse with the patient, without the

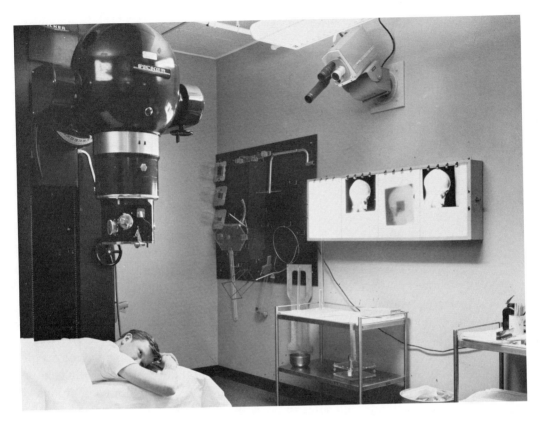

COURTESY COHU ELECTRONICS, INC.

FIGURE 1-9. A television camera monitors patients undergoing radiation treatment.

inconvenience and expense involved in a trip by either. The time thus saved allows the doctor to see additional patients. The primary difficulties with such a system have generally been associated with the narrow bandwidths that were dictated by long distance transmission. However, recent developments in wideband microwave transmission equipment have made such installations practical and extremely attractive.

Television, especially color television, can be an able assistant in many specialized areas of medicine. For example, when a television camera is used in conjunction with a cystoscope and a suitable light source, pictures obtained from within the body can be quite useful in diagnosis.[1] Blood

[1] Bush, Wilkey, Meyer and Brandy, "Uses of Television Cystoscopy," *Journal of the SMPTE,* Volume 76, Nov., 1967.

vessel patterns are easily discerned and various types of eruptions and inflammations may be observed.

In medical research, specialized cameras may be of use. For instance, television cameras employing vidicon tubes which have their greatest sensitivity in the ultraviolet region of the spectrum have been used in cancer research. Tubes are also available which are sensitive to infrared radiation, opening other broad areas in detection of localized infections.

Other uses, too numerous to mention, indicate the expanding potential of television in medical applications.

## SURVEILLANCE

Probably the most widely used single application of closed circuit television is that of surveillance. Stores and businesses of all kinds use the television camera to keep an eye on their stock, entrances, warehouses and parking lots. Banks and financial institutions use the television camera for security reasons. Manufacturers monitor assembly line status with television and law enforcement agencies can use it for crowd and building surveillance.

Possibly the biggest problem associated with this application is providing sufficient light for proper camera operation. During daylight hours, especially if the camera is mounted out of doors, the lighting conditions are usually adequate, but after dark, additional light must be provided. Most surveillance cameras use a vidicon tube as the light-sensitive element, but it will generally not provide a usable picture with scene illumination below approximately one to five foot-candles, depending upon the particular tube used. There are low light-level cameras that can provide images well below this illumination level, but they are quite specialized and expensive.

It is usually desirable that cameras which are to operate unattended be equipped with some sort of automatic sensitivity control circuitry. Cameras thus equipped automatically compensate for changes in light level on the scene, usually by adjusting voltages at the camera pickup tube. Such compensation may remain effective over light variations of up to 10,000 to 1. This provides a useful picture with illumination levels that range from bright sunlight, to that provided by street lights or floodlights during nighttime operation.

There are other, more specialized, surveillance applications where a television camera is extremely helpful. For instance, Figure 1-10 shows the control board in a power generating plant, where two monitors are display-

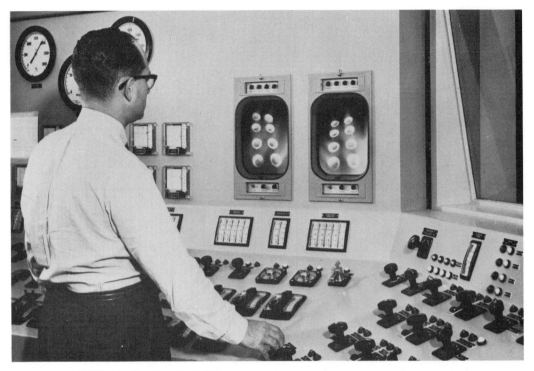

COURTESY COHU ELECTRONICS, INC.

FIGURE 1-10. Cameras monitor boiler flames in a power generating plant.

ing images of the gas flames used to fire the boilers. The cameras used in this instance are located directly at the furnace viewing ports and, because of the extreme heat in these areas, must be externally cooled by water jackets or circulated air to prevent damage to the electrical components. Such monitoring provides instant recognition of any change in the flame pattern and corrective measures can be made accordingly.

## DATA TRANSMISSION

The use of television cameras for data transmission is becoming increasingly popular as the need for instant communication expands. Banks use it to approve checks from a remote location and cameras located at unattended plant gates can check identification cards and individuals before the gate is opened by remote control.

Instant transmission of documents, letters, and photographs (Figure

1-11) can be routed to one or many destinations, providing pictoral information instantly. Without television, dissemination of the same information would be a lengthy and costly process.

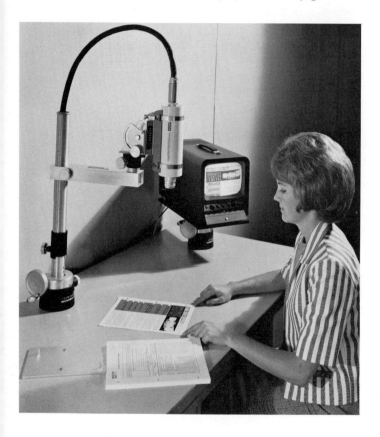

COURTESY COHU ELECTRONICS, INC.

FIGURE 1-11. High-resolution television cameras speed data transmission.

The primary consideration in choosing a television camera for data transmission must be the camera's ability to resolve detail. This, in turn, is dictated by the type of information that the user wishes to televise. A standard typewritten page is often chosen as a criterion, and to be truly legible at the monitor, high bandwidths and scanning ratios are needed. Of course, at times resolution elements in a document will fall below the size of a typewritten character. In these instances, it is important to utilize cameras that are indeed "high-resolution" devices.

The versatility that such systems lend to data transmission is indicated in Figure 1-12. By using a video switcher to select any of the video signals from the cameras shown, a user (seated at his desk at a remote location) could have immediately available the information coming in on any of the

COURTESY COHU ELECTRONICS, INC.

FIGURE 1-12. JPL "Space Command" center speeds information flow with television cameras.

15 teletypes shown. If this same information is made available to a large number of other interested parties for selective viewing, the system becomes an even more remarkable communication device.

## RADIATION

The television camera has achieved an added responsibility with the advent of nuclear reactors and radioactive materials. Televised images of the interior of reactors during the process of refueling make such cameras an extremely important part of the total operation. Indeed, the ability to restore a reactor to proper operation quickly often depends upon the camera's ability to operate reliably in high radiation areas.

Cameras designed for high radiation applications do not normally employ semiconductor devices in the camera head because their operation is seriously affected by the radiation. Also, other materials, normally used in the construction of television cameras, experience a change in chemical composition because of gamma radiation and, in the case of teflon (which is often used as an insulator), will harden and even powder.

Camera housing materials must be carefully chosen to be impervious to the effects of radiation, and should not become radioactive carriers themselves after periods of exposure. Special stainless steel materials, containing no materials with appreciable half-lives, are used for this purpose. Stainless steel also allows the camera to be decontaminated by being dipped or washed in solutions such as dilute nitric acid. Of course, such a camera must be hermetically sealed to preclude the admission of radioactive dust or acid components, and the housing material is often highly polished to prevent contaminant retention in small pits or scratches in the surface.

Vacuum tubes, nuvistors, etc. are usually used as the active elements in the radiation camera head. Since the photosensitive device (usually a vidicon) is also a vacuum tube, its operation is not greatly affected by radiation. However, if the camera is to operate in areas of intense radiation, the optical elements must be made of a specially composed glass. Ordinary optical glass turns brown with sufficient exposure to radiation and a resulting decrease in the speed or sensitivity of the optical system is experienced. These elements are therefore normally constructed of special nonbrowning glass.

Figure 1-13 shows a radiation camera head equipped with a mirror attachment for right-angle viewing within the narrow confines of a reactor duct. Small lights may be contained within such housings for illumination.

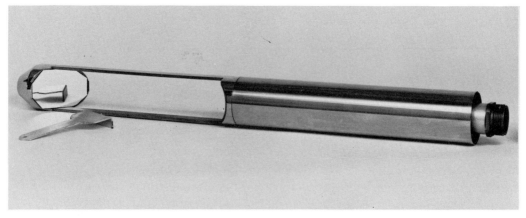

COURTESY COHU ELECTRONICS, INC.

FIGURE 1-13. Radiation-resistant cameras are used to view reactor ducts.

## UNDERWATER

The underwater television camera is generally a miniaturized closed circuit camera enclosed in a much more sophisticated and durable housing. Cameras operating at extreme depths must be able to withstand pressures of several tons per square inch without allowing stress or moisture to be transmitted to the camera circuitry or optics. Cameras that are to operate in a salt water environment must also be constructed of materials which withstand the corrosive effects of electrolysis. It is equally important to electrically isolate the camera from the housing, otherwise the effects of electrolysis will be accelerated. Stainless steel is one of the best materials to use, and may be the only suitable material for many applications. In applications of moderate depth, where cost is a major factor, aluminum (hard anodized) may be used.

Good lighting is essential to effective video reproduction of the underwater scene. Due to the dense medium and the many impurities that may be present (as well as the absence of natural light at great depths) intense light sources are demanded by many applications. These sources must be able to withstand the same pressures as does the camera system in Figure 1-14.

In conjunction with proper lighting, care must be exercised in selecting

COURTESY COHU ELECTRONICS, INC.

FIGURE 1-14. Underwater television cameras may be used to great depths.

a proper lens. Wide angle lenses are normally used, since most objects viewed by an underwater camera are close to the lens (as dictated by available light, visibility, etc.). Lenses of long focal length normally will not focus at short distances and their narrow angle of view limits their usefulness in underwater applications.

The viewing angle of most television cameras in the underwater environment is approximately 70% of that of the same camera in air, due to the different index of refraction (see Chapter Three) encountered in the submerged condition. Objects underwater appear to be some 30% nearer than they actually are. Special optically corrected faceplates may be used to eliminate this distortion.

Underwater systems may be generally classified by depth capability into three groups. These are:

### Shallow

This includes systems operating to depths of 600 feet (100 fathoms) or less. Included might be those systems designed for sewer inspection, etc. that are watertight but not expected to withstand pressure equal to more than a few feet of water. Underwater operation in harbors, rivers, and small lakes also utilizes systems designed for shallow depths.

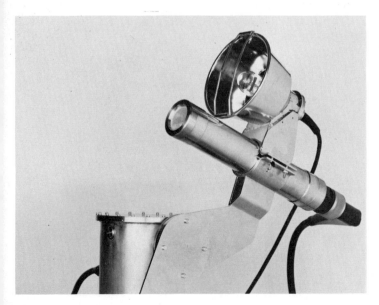

COURTESY COHU ELECTRONICS, INC.

FIGURE 1-15. Systems completely housed in stainless steel resist the corrosive effects of sea water.

### Moderate

Systems designed for moderate depths are generally employed for use with underwater oil drilling, bottom surveys, etc. They are usually designed to operate to that depth constituting the continental shelves of the large land, masses, and are seldom submerged beyond 1500 feet.

### Deep

TV cameras designed for depths beyond 1500 feet (Figure 1-15) are used for scientific exploration and ordnance retrieval.

With the increasing interest in developing the vast resources of the oceans, it is inevitable that the television camera will play in important role.

## AIRBORNE

Television cameras are being increasingly utilized in a wide variety of airborne applications. While motion picture photography has long been used to observe various aircraft functions and armament tests during flight, in many of these applications the pilot has no way of determining just what he is recording on film. A television camera mounted beside the film unit (Figure 1-16) largely eliminates the problem. The pilot simply watches

COURTESY U.S. AIR FORCE.

FIGURE 1-16. A television camera (right) allows the aircraft pilot to view what the film camera (left) is filming.

a small monitor in the cockpit and keeps the aircraft oriented to maintain the target in the center of the screen, thus assuring himself of a good photographic run. Such systems are widely used, particularly with the military, in areas where high speed, single-seated aircraft are dictated.

Airborne systems are also being used increasingly in drone (unmanned) aircraft. The wideband transmission systems that have become available, permit excellent resolution capabilities, creating applications in instrumentation monitoring, reconnaissance, and other airborne techniques. Manned aircraft may also find specific needs for on-board monitoring of critical events (Figure 1-17).

There are several problems associated with the use of television systems in the airborne environment. The first is vibration, which can cause effects such as camera tube microphonics, broken or dislodged electrical components, and intermittent connections. The second problem that arises

COURTESY U.S. AIR FORCE.

FIGURE 1-17. Exterior cameras (circled) allow continuous monitoring of X-15 rocket ship prior to launching.

is the effect of high altitude on the high voltages that are used in association with vidicon and kinescope tubes employed. Arcing and corona problems become especially troublesome between altitudes of 80,000 to 120,000 feet, depending also upon other factors such as humidity, voltage potential, etc.

Another problem that occasionally occurs in high altitude aircraft is the undesirable effect of extreme cold at great heights on the camera circuitry and the photosensitive surface of a vidicon tube. This may be overcome by placing the camera within the main body of the aircraft where temperatures are not so extreme, or providing a means of heating the camera (Figure 1-18).

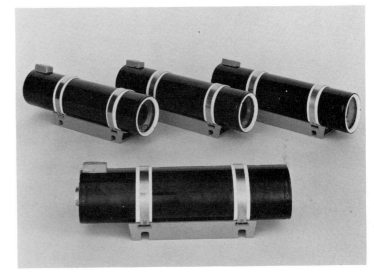

COURTESY COHU ELECTRONICS, INC.

FIGURE 1-18. Camera housings encased in heater jackets assure operation in extreme cold.

## SPACE

One of the most exciting and noteworthy uses of the television camera has been its use in various space programs. It provided man's first look at the rear surface of the moon, and relayed literally thousands of excellent photographs from the moon's surface. Cameras mounted in weather satellites are continuously photographing the cloud cover that moves across the earth, and radio transmitters relay the pictures to ground stations where they provide valuable insight into the earth's weather patterns.

Most of the pictures obtained from spacecraft are a product of slow-scan television systems such as that employed aboard the "Surveyor" spacecraft series (Figure 1-19). The camera, shown in cross-section in Figure

COURTESY JET PROPULSION LABORATORY.

FIGURE 1-19. The Surveyor VII spacecraft televised the moon's surface with exceptional clarity.

1-20, operates on the principle that its image tube stores scene information until scanned by an electron beam. A shutter assembly, similar to that of the familiar family camera is used, allowing a brief exposure to the scene. A slow moving beam of electrons then converts the scene into relatively low frequency electrical signals, effecting considerable bandwidth savings. Such low frequency signals are also easier to separate from the noise that interferes with radio transmissions from space.

The Surveyor camera observed the lunar landscape through a mirror assembly that could be caused to rotate and tilt, allowing a view of large areas surrounding the spacecraft. Figure 1-21 is a spherical mosaic of pictures taken by Surveyor VII's television camera. Adding the smaller

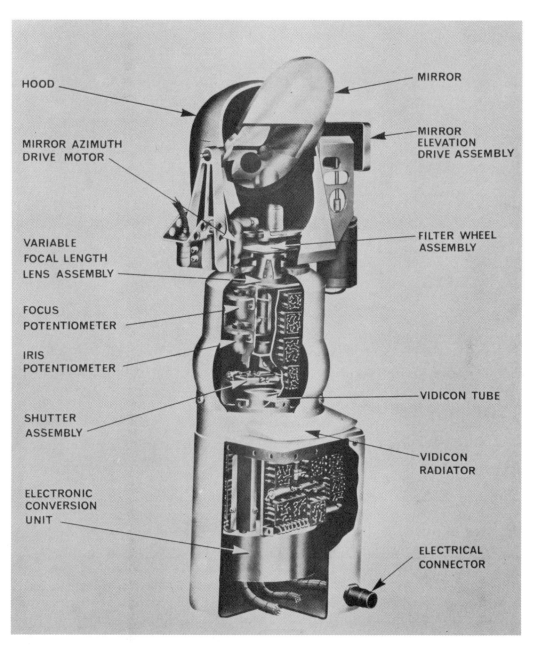

HOOD

MIRROR AZIMUTH
DRIVE MOTOR

VARIABLE
FOCAL LENGTH
LENS ASSEMBLY

FOCUS
POTENTIOMETER

IRIS
POTENTIOMETER

SHUTTER
ASSEMBLY

ELECTRONIC
CONVERSION
UNIT

MIRROR

MIRROR
ELEVATION
DRIVE ASSEMBLY

FILTER WHEEL
ASSEMBLY

VIDICON TUBE

VIDICON
RADIATOR

ELECTRICAL
CONNECTOR

COURTESY JET PROPULSION LABORATORY.

**FIGURE** 1-20. An exploded view of Surveyor VII's on-board television camera.

COURTESY JET PROPULSION LABORATORY.

FIGURE 1-21. The rock-strewn moonscape about 18 miles north of the crater Tycho as reconstructed from televised images from Surveyor VII.

individual pictures into a composite affords a panoramic view of the lunar terrain that is extremely valuable to scientists evaluating the structure and composition of the surface material.

Another historic photograph is shown in Figure 1-22. This picture was obtained using a slow-scan camera aboard the Mariner IV spacecraft as it passed the planet Mars at a slant range of about 7800 miles. The series of photographs obtained on this mission constituted the first closeup glimpses that man obtained of any of the earth's neighboring planets.

The role of the television camera in space will be increasingly important. Television cameras mounted aboard robot spacecraft will continue to scout future destinations of manned vehicles and send back pictures of what lies ahead. Improvements now being made in camera pickup tubes, integrated circuits, and power supplies, will result in cameras with greatly increased capabilities. The potentialities of television in this rather awesome application have only been sampled.

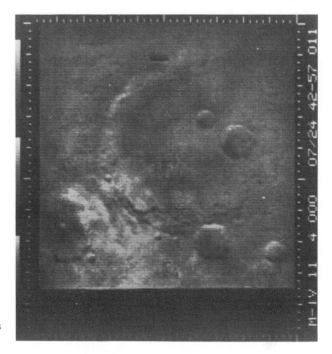

FIGURE 1-22. The surface of Mars as seen by Mariner IV spacecraft.

## OTHER APPLICATIONS

Certain television camera pickup tubes possess the ability to react to light that lies outside the visible spectrum. Some types can, for example, "see" the radiant infrared frequencies emitted by warm objects. Use of a camera employing such a device would enable army personnel to visually detect an enemy vehicle at night by the heat of its engine. Similarly, a normally invisible laser beam can be observed directly on a television monitor.

Figure 1-23 illustrates the results obtained using a tube sensitive to infrared radiation in a standard television camera. The two monitors are displaying the outputs of two cameras, the one on the left showing video obtained using a standard vidicon tube, and the right monitor presenting the output of the infrared sensitive tube. Both cameras are aimed at a jet of burning hydrogen that is normally not visible. The monitor on the left displays the scene as the human eye would see it, but the right-hand monitor clearly shows the flame. Infrared radiation is emitted by water molecules produced by hydrogen burning with air, or pure oxygen, and

FIGURE 1-23. A hydrogen flame as seen by a standard (left) and an infrared sensitive (right) television camera.

the radiation is picked up by the camera tube designed for infrared detection. The infrared tube also has just enough sensitivity to the visible spectrum to show the surroundings as reference.

Camera tubes that are sensitive to ultraviolet light are also available, but are not widely used. Normal optical systems are relatively opaque to light in this portion of the spectrum. However, they are used to a limited extent to observe biological specimens being illuminated by relatively high levels of ultraviolet light.

Another application of television cameras that excites the imagination is their use to see in areas where the available light is so low that the unaided human eye sees only darkness. Such specialized cameras normally employ *image intensifiers* to amplify the light image before viewing it with the photosensitive element of the camera tube. Military applications are immediately evident.

The list of television applications could go on almost indefinitely. Their tasks range from the menial to the memorable.

The television camera is truly a remarkable device.

# BASIC TELEVISION
# SYSTEMS

A television system may consist of a simple camera connected to a television receiver, or it can be an expensive complex of highly sophisticated electronic equipment geared to the needs of a broadcast network or a space program. Whatever the application, the television system should always be tailored to the needs, both present and future, of the user but should not be unnecessarily extravagant.

## THE TELEVISION CAMERA

Television cameras may assume many different physical and electrical configurations. However, in general, they may be divided into two basic groups—self-contained cameras and two-unit systems that employ separate camera heads driven by remotely located camera control units.

The self-contained camera (Figure 2-1) contains all of the elements necessary to view a scene and generate a complete television signal. It is only necessary to supply power (usually ac) to the unit and provide a means for routing the video signal to a monitor. All power supplies, scanning circuits, and video amplifiers are contained within the unit. Such cameras may also incorporate a sync generator, or have the capability of being driven by a remotely located master sync generator.

Figure 2-2 illustrates the two-unit concept of a television camera that has become so popular with industrial and scientific users of closed circuit systems. The camera head and its associated camera control unit may

COURTESY GPL.

FIGURE 2-1. A self-contained television camera.

COURTESY GPL.

FIGURE 2-2. A television camera head with its associated camera control unit.

contain essentially the same electrical components as the self-contained camera, but the bulk of the circuitry is usually contained in the camera control unit, which is then connected to the camera head by means of a multiconductor cable. Thus, the camera head can be placed in locations where heat, dust, shock, vibration, etc. may be rather severe, without endangering the major portion of the circuitry which is located in the camera control unit. When the remote camera head is used, it also has the advantage of being smaller than a self-contained unit would be.

The remote camera head usually contains only the photosensitive pickup tube and its associated deflection yoke, a video preamplifier, and some filtering and deflection circuits. Some cameras may also employ small motors to provide a means to remotely focus the lens (or move the camera tube back and forth to obtain proper focus). Exact configuration will depend upon the application and the manufacturer of the equipment.

When a remote camera control unit is used, most of the electrical operating and setup controls are contained within it. For this reason, the camera control is usually mounted near a viewing monitor so the results of any adjustments may be easily viewed on the monitor screen. The camera head will require little or no adjustment since it may be located in a relatively inaccessible area. It is sometimes desirable to house the circuitry of more than one camera control within the same chassis (Figure 2-3).

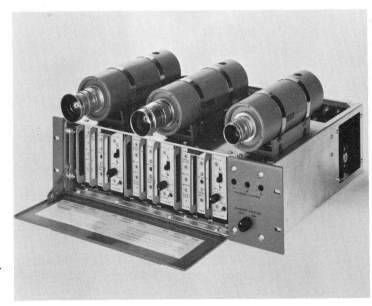

FIGURE 2-3. A multiunit camera control that drives three camera heads.

The cable that typically connects the camera and camera control will carry the control and high voltages necessary for the camera tube, as well as video, deflection, and other signals. Its length may vary from a few feet to several thousand feet, although lengths in excess of 2000 feet are rare.

Television cameras may also be classified according to resolution capabilities, bandwidth, scanning ratio, environmental abilities, and the like. These and other parameters will be covered in detail in following chapters.

## SYNCHRONIZING SYSTEMS

To generate a meaningful presentation on the raster of a monitor or receiver, some means is needed to synchronize the scanning systems of both the camera and the monitor. Many cameras have synchronizing (sync) generators built right into them. Such units may vary considerably in configuration, and in some closed circuit television cameras may generate only a rudimentary signal for use at the monitor. However, when a stable picture that equals or exceeds broadcast quality is desired, it is generally necessary to employ a somewhat more sophisticated device, either within the camera or as an accessory unit. Such a unit develops signals necessary to drive the deflection systems in the camera and, at the same time, inserts a waveform onto the video output signal that will cause the monitor scanning system to "slave" to the camera.

When several cameras are to be used, it is often a good idea to have them all synchronized by a single sync generator. This will many times be more economical, since it then becomes unnecessary to include one in every camera. It is also advantageous because all cameras are scanning "in time" with one another instead of each operating with individually generated synchronizing signals. Broadcasting applications provide a good example of synchronous multicamera systems. When the scene is shifted from one camera to another, the synchronizing waveforms are in phase, so the monitor or home receiver is not interrupted in its scanning process. However, if each camera had its own sync system, when switching from one camera to another, the monitor or receiver would have to readjust its scanning procedure for each camera and the picture might "roll" momentarily while this was being accomplished.

Figure 2-4 illustrates one method of driving multiple cameras from a single sync generator. The outputs from the generator (here shown as a single line, labeled "sync") go directly to camera 1, where they are

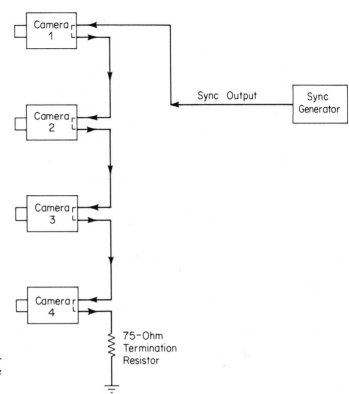

FIGURE 2-4. "Loop-through" connections allow a sync generator to drive several camera systems.

connected in a "loop through" fashion. When the line goes into the camera, the sync waveforms are sampled by circuits which have such high impedance that they do not affect the waveform. The camera uses this sample waveform to develop drive signals for the deflection oscillators. The sync is then routed back out of camera 1 to camera 2, where it is again sampled and looped out to camera 3. When the sync is looped out of camera 4, it is terminated by a resistor connected to ground. The terminating resistor is generally a 75-ohm resistor because the output of most television camera equipment is designed to work into a 75-ohm load. In all cases the termination must occur at the extreme end of the line or the waveforms may become severely distorted.

## MONITORS AND RECEIVERS

The television receiver is a familiar part of a television system. The home receiver is seldom thought of as a part of a system, yet in a very real

COURTESY COHU ELECTRONICS, INC.

FIGURE 2-5. A closed circuit television monitor.

sense it is the most important part of the largest and most complicated television system in the world. The receiver may also be called a "monitor," but common terminology associates the term monitor with a closed circuit or studio television display device that contains no provisions for receiving broadcast signals, but rather relies upon a direct input of unmodulated video signal. Monitors are usually associated with closed circuit television and receivers with broadcast television.

Television receivers in the United States are all designed to operate at a fixed scanning ratio and generally conform to the same approximate bandwidth and resolution capabilities. Not so the television monitor (Figure 2-5), which is often used at different scan rates and may exhibit widely varying bandwidth and resolution capabilities. Because closed circuit monitors are "hard-wired" to the camera systems, the bandwidth need not be limited to conform to federal broadcast regulations, and many monitors therefore have bandwidths that extend beyond 30 megahertz.

Several monitors may be used to display the scene being viewed by a single camera, as shown in Figure 2-6. As was the case with the previous example of sync distribution, video may be looped through a series of monitors, with the last of the series terminated in a 75-ohm resistor. Most

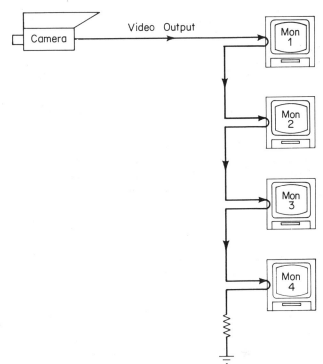

FIGURE 2-6. "Loop-through" connections allow several monitors to display
the output of a single camera.

monitors are equipped with a termination switch that terminates the input video line with a 75-ohm resistor in one position and leaves the input un-terminated in the other position.

When the monitors in the above example are to be operated at con-siderable distances from one another, it may be impractical to connect them in a series arrangement. In such cases it is often necessary to employ a *distribution amplifier* to route the video properly.

## DISTRIBUTION AMPLIFIERS

A distribution amplifier functions to convert a single signal line into multiple distribution paths without excessively loading or distorting the original signal. Figure 2-7 shows a four-output amplifier being used to drive four terminated monitors. The input to the distribution amplifier must, of course, be terminated in 75 ohms to provide proper loading for the camera output. Looping inputs may also be incorporated on distribu-

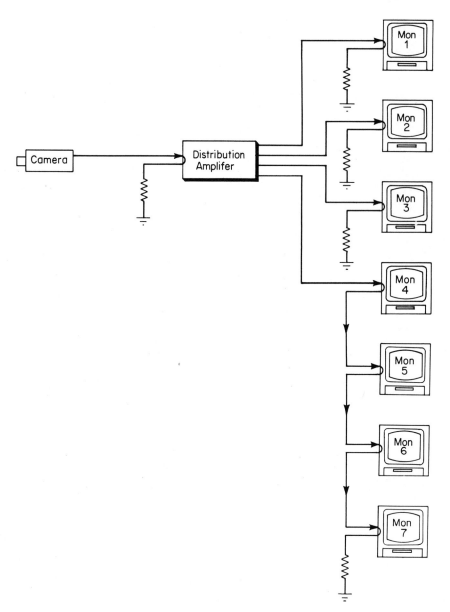

FIGURE 2-7. Distribution amplifier allows more versatility in signal routing.

tion amplifiers to allow the addition of more amplifiers to provide as many feed lines as desired. Again, the last such unit would be the one terminated.

All outputs from a distribution amplifier must be properly terminated to provide correct operation. However, since all outputs are essentially identical to the input, any of them may be used to drive a series string of monitors, as shown in the diagram. Many configurations of such distribution amplifier systems are possible.

Distribution amplifiers are generally classified into two general categories—video distribution amplifiers (VDA's) and pulse distribution amplifiers (PDA's). The PDA's are used to route signals from sync generators and other similar equipment which provide pulse waveforms having fast rise and fall times.

Distribution amplifiers are generally used only when there are many using destinations, or the destinations are too widely separated to allow series looping connections.

## VIDEO SWITCHERS

It is very often desirable to provide a means for viewing the output of several cameras on one monitor. This is simply done by switching the monitor input from one camera output to another, as shown in Figure 2-8. By means of a simple toggle switch, the operator can then select the desired scene and have the capability of transferring to another at the desired moment. (It should be noted that proper termination resistors for the cameras are provided whether or not they are being viewed.) Simple switches such as this have many advantages. They are very small, inexpensive, require no power, and are extremely reliable and simple to maintain.

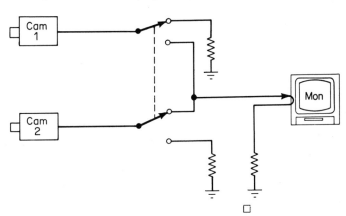

FIGURE 2-8. A simple video switching circuit.

Pushbutton operated video switchers are more desirable from an operator standpoint. A simple industrial type system is shown in Figure 2-9. The circled crosspoints indicate a pushbutton switch which, when depressed, connects the associated camera to the monitor. The switches not depressed connect the terminating resistors to the appropriate camera. In operation, all switches are interlocked so that only one camera can be connected to the monitor at any time. Depressing one switch releases all others. Thus, various camera outputs can be switched onto the monitor by simply depressing the proper switches in a desired sequence.

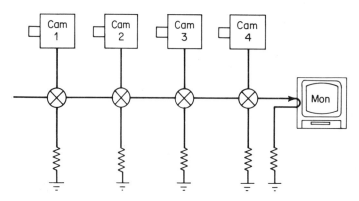

FIGURE 2-9. Another video switching system. Circled X's indicate pushbutton switches.

## REMOTE CONTROLS

There are several types of remote controls associated with television equipment. The elements of lenses can be motor driven and remotely controlled. It may also be desirable to exercise remote control over various camera functions, even if the camera may be inherently self-contained. Operators may wish to adjust parameters such as video gain, camera sensitivity, blanking level, video polarity, and the like. This is especially true in an educational or broadcast situation where the video signal must be maintained within very close tolerances of amplitude.

In many applications, primarily closed circuit, it is desirable to be able to move the camera *remotely* up and down and around its central axis to view different sections of a scene. In such cases, pan-tilt units (Figure 2-10) may be used to position the camera remotely. These units typically provide a 360-degree rotational capability and allow tilting action to plus-or-minus 90 degrees.

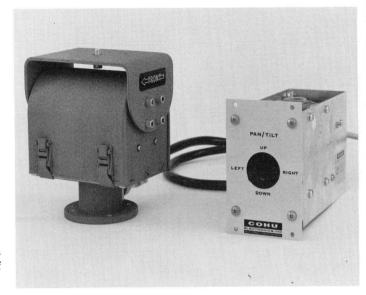

FIGURE 2-10. A pan-tilt unit and its remote control allow positioning of cameras by distant operators.

## OTHER EQUIPMENT

There is a wealth of additional equipment that may be used as a part of, or in conjunction with, a television system. Microphones, audio amplifiers, and distribution systems might be incorporated. Video tape recorders might also be used for the added versatility they can add to a system. Microwave relay links may also be demanded in certain applications and RF modulators could find use in institutions, such as hospitals, schools, etc., where it is desirable to utilize the output from a camera to display information on standard UHF/VHF receivers. Also, lighting equipment may be a necessary part of some television systems.

FIGURE 2-11. Proper combinations of equipment result in a versatile system.

# PRINCIPLES
# OF OPTICS

To televise an image it is first necessary to process the light emitted or reflected from a scene so that an accurate image of the scene is formed on the photosensitive surface of the camera tube. This is accomplished by the use of a lens chosen to fulfill the requirements of a particular application.

Although a person involved with the use of television systems need not be an expert in the field of optics, a fundamental knowledge of the principles involved can be extremely helpful. Intelligent selection of lenses can greatly improve the quality of the final televised image.

Lenses rely upon characteristics exhibited by light when it contacts or traverses a medium whose density is materially different from that of the original medium. Knowledge of these characteristics and the practical manner in which they are utilized lend insight into the basic essentials of lens theory.

## REFLECTION

The human eye is able to see objects because they reflect light. Most objects, because they are so coarsely textured with respect to the wavelengths of light, reflect in many directions. This tends to diffuse the light and make the object visible from all angles. However, when a finely textured or highly polished surface is encountered by light, the reflected light is not diffused but maintains a definite directional characteristic.

A flashlight aimed at a mirror is a good example of the reflection capabilities of a smooth surface. The mirror changes the direction of the beam but does little to alter its dimensional properties (Figure 3-1). For this reason mirrors are sometimes used to collect, process, and direct light in a lens system. It is interesting to note from the diagram that the angle at which light is reflected from the flat surface is always exactly equal to the angle of the original light beam, using the plane of the mirror as reference.

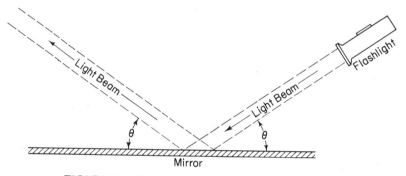

FIGURE 3-1. Reflected light from a plane surface assumes an angle with the surface equal to that of the original light beam.

A mirror that is spherically curved may be used to form an image. Figure 3-2 illustrates a classic example of image formation using an arrow as the object. It can be seen that the spherical nature of the mirror causes the light from the tip of the arrow to be reflected to a single point, forming an image of the tip. Light from other points along the length of the arrow are similarly reflected, with the base of the arrow reflecting back along its original path to form the base of the image. The point at which the image

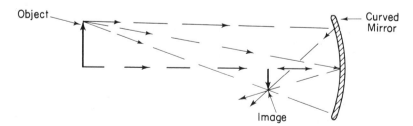

FIGURE 3-2. A spherically curved mirror may be used to form an image.

is formed is a function of the object-to-mirror distance and the radius of curvature of the mirror. Images thus produced are always inverted.

Imaging by use of mirrors is not normally practiced in lenses used with television cameras, but telephoto lenses of extreme focal length sometimes utilize mirrors to good effect (Figure 3-3).

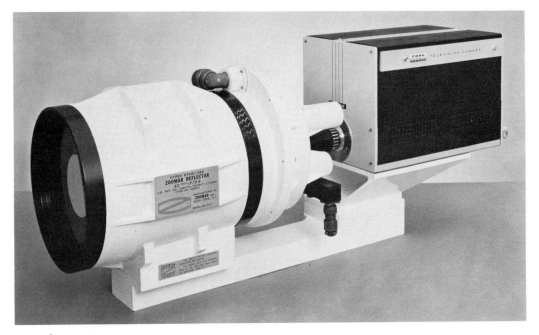

COURTESY ZOOMAR, INC.

FIGURE 3-3. A 40-inch focal length lens uses mirrors to achieve proper imaging within the confines of a relatively short housing.

## REFRACTION

Most people have observed that a straight object, such as a pole or boat oar, appears to bend when it is partially immersed in water. This phenomonon is generally dismissed in a casual manner but, properly utilized, it plays an important role in the construction of lenses. The effect is caused because the speed of light varies with respect to the density of the transparent medium in which it is traveling. To understand more easily, it is convenient to illustrate the incident light as an advancing plane, or wavefront, as shown in Figure 3-4. Here the light is entering a trans-

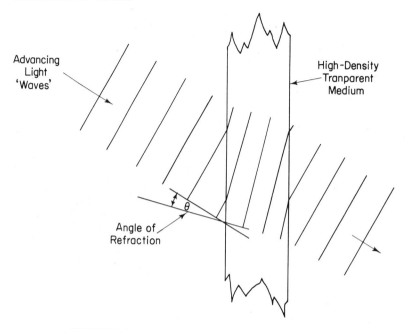

Advancing
Light
'Waves'

High-Density
Tranparent
Medium

θ

Angle of
Refraction

FIGURE 3-4. Refraction alters the directional characteristic of light.

parent medium, such as glass, whose density is greater than the air above it. Since the velocity of light is less in the denser medium, that portion of the wavefront which has passed beyond the surface has slowed, causing an effective change in the direction of travel. This effect is known as *refraction* and the extent to which it occurs is dependent upon a property of the two mediums which is termed the *index of refraction*. The index of refraction is a ratio of the velocity of light in the particular medium under investigation compared to the speed of light in a vacuum. For all practical purposes, the speed of light in air is considered to be the same as it is in a vacuum, giving air an *index* of 1.0. The index of water is about 1.3 and optical-type glass usually has an index of approximately 1.5 to 1.9.

Using parallel beams of light as an illustration, Figure 3-5 shows the effect of passing an angular beam of light through a piece of plate glass. Upon entry into the glass, the angle changes because of the change in index, as previously discussed. As the light emerges from the glass, the original angle is restored because the change in index is essentially the reverse of that encountered when the light entered the glass. The offset experienced by the beam is a function not only of the index of refraction, but the thickness of the glass and the angle of entry as well. This action

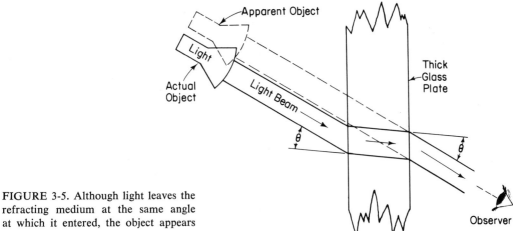

FIGURE 3-5. Although light leaves the refracting medium at the same angle at which it entered, the object appears displaced.

becomes quite important in the consideration of lenses that use the refractive principle as a means of image formation.

Figure 3-6 depicts refraction through a piece of glass whose surfaces have been spherically formed. The glass thickness is obviously not constant across the diameter of the lens. Consequently, those rays that pass through the center will be slowed for a longer period of time than those that pass near the edge. The entry angle of light into the glass will also be varied because of the curvature. The simple lens thus constructed has the ability to form an image as illustrated. In this case the object is assumed to be a point source of light located on the axis of the lens. Rays emitted by the object $O$ enter the lens at various diverging angles but, due to the spherical nature of the lens surfaces, the refractive action causes them to converge at a distance $F$ on the opposite side of the lens. The point at which the image is formed is called the *focal point* or, in instances where

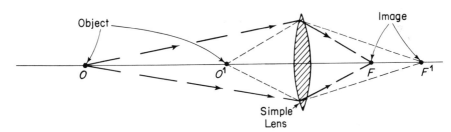

FIGURE 3-6. Refraction in a simple lens produces an image. The distance at which the image is formed depends upon object distance and the material and curvature of the lens.

the image covers a larger area, the *focal plane*. Object *O'* is located nearer to the lens and the rays that reach the lens from this source are more divergent in nature, resulting in a focal point that is farther from the lens. Thus, *the focal point is a function of the distance of the object from the lens*. This effect is most pronounced when objects are relatively close to the lens. At close distances a small change in distance from the lens results in a rather sizable change in divergence angle of the light rays passing through the lens. In contrast to this, the rays reaching a lens from a point a hundred feet away are little different, with respect to divergence angle, than those coming from a point at infinity, and the focal point for either is almost identical. At these longer distances the divergence angle is so small that the rays are essentially parallel (Figure 3-7).

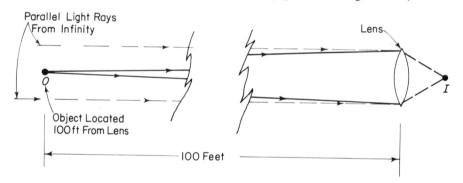

FIGURE 3-7. For most simple lenses of short focal length an object at infinity and one at 100 feet or more will share a common focal point.

## FOCAL LENGTH

The *focal length* of a lens is the distance between the center of a lens and its focal point *when the object is at infinity*. The object distance prerequisite assures that all incoming rays will be parallel. Therefore, a lens that forms an image of a far distant object at a point one-half inch behind its optical center is said to have a focal length of one-half inch. Lenses of longer focal length are, of course, physically greater in length and offer an effective magnification that is greater than that obtained from lenses of shorter focal length. For instance, a six-inch lens has a magnification capability six times greater than that of a one-inch lens. In television systems that use vidicon tubes (see Chapter Four) at the focal plane, it is general practice to express the magnification of a lens by using a one-inch focal length lens as a reference. Therefore, a one-half-inch lens has only a 0.5:1 magnification, or half as much as a one-inch lens.

Figure 3-8 illustrates the effect of differing focal lengths on the size of an image, using a one-inch and a two-inch lens as examples. When the object is located the same distance away for both lenses, the image of the two-inch lens is twice the size of that formed by the lens with a one-inch focal length. The object in this instance is no longer a point source of light, and it becomes evident that an image formed by a refractive lens is inverted, such as that formed by the reflector previously examined. If the focal plane upon which the image is focused is the faceplate of a camera tube (shown in heavy dotted lines) the difference in magnification of the two lenses is immediately apparent. A lens with a three-inch focal length would enlarge the image even further, to the extent that a portion of it would fall outside the limits of the camera tube faceplate. Therefore, when it is considered necessary to examine distant objects in detail, a lens with a long focal length should be chosen.

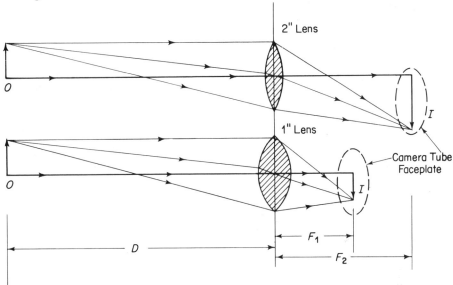

FIGURE 3-8. Lenses of different focal length affect the size of the image.

Lens focal lengths are usually expressed in either inches or millimeters. To convert from one to the other, the following simple equation applies:

$$\frac{\text{millimeters}}{25.4} = \text{inches}$$

Thus, a 25 mm lens is approximately the same as a 1-inch lens, a 50-millimeter lens is the equivalent of a 2-inch lens, and so on.

Almost all television cameras employ lenses whose focal length can be slightly altered (Figure 3-9). This is accomplished by incorporating a means to move the front element of the lens, and generally takes the form of a "focus ring" located on the barrel of the lens. This important addition allows sharp focusing on objects that may be located at various distances from the lens. The focus element on most lenses is calibrated in feet, and by setting it to five feet, for example, an object located five feet from the lens will be brought into sharp focus. Additional aspects of focusing will be discussed in the section "Depth of Field."

FIGURE 3-9. Complex lenses of various focal length. Adjustments allow focusing and control the amount of light that passes.

## VIEWING ANGLE AND FIELD OF VIEW

Referring again to Figure 3-8, it can be seen that a one-inch lens develops an image of the arrow that is smaller than that produced by a two-inch lens, but the one-inch lens "sees" more of the area that surrounds the arrow. Restated, this means that a one-inch lens has a wider *viewing angle* than a two-inch lens.

There is an irrevocable relationship between lens focal length and viewing angle. Lenses of short focal length are said to be *wide-angle* lenses and those with long focal length are termed *narrow-angle*. When employed on television cameras using vidicon pickup tubes these two terms are usually referenced to a one-inch lens, which is considered to be the normal. Common usage has made these terms rather broad in inter-pretation.

Viewing angle of a television lens is normally correlated to the por-tion of the focused image which actually falls upon the useful area of the camera pickup tube. In the case of a vidicon tube this area is ½ inch wide by ⅜ inch high. Consequently, the vertical angle of view, as finally seen on a monitor, will be limited to ¾ that of the horizontal. Table 3-1 lists viewing angles obtained from some common lenses designed for use with vidicon television cameras.

**Table 3-1**

VIEWING ANGLES OF TYPICAL
VIDICON CAMERA LENSES

| Lens<br>Focal Length<br>(inches) | Horizontal<br>View Angle<br>(degrees) | View Angle<br>Vertical<br>(degrees) |
|:---:|:---:|:---:|
| 1/2 | 53 | 39 |
| 1 | 28 | 21 |
| 2 | 14 | 10.5 |
| 3 | 9.6 | 7.2 |
| 4 | 7.1 | 5.3 |
| 6 | 4.8 | 3.6 |
| 12 | 2.2 | 1.7 |

Closely associated with the viewing angle of a lens is *field of view*. Field of view describes the width and height of the scene viewed and is determined by lens focal length (viewing angle) and the lens-to-subject distance. Since the useful image on the pickup tube is rectangular in nature, it is common to state the field of view in terms of both width and height.

Field of view for a vidicon camera lens can be quickly determined by use of the simple equation:

$$W = \frac{D}{2 \times F}$$

where

$W$ = width of field
$D$ = lens-to-subject distance
$F$ = lens focal length (in inches)
$2$ = ½-inch width of vidicon

(Note: $W$ and $D$ must be expressed in the same units: inches, feet, meters, etc.) Therefore, if distance to the subject is assumed to be 20 feet and the lens in use has a two-inch focal length, the width of field will be:

$$W = \frac{20 \text{ feet}}{2 \times 2 \text{ inches}} = 5 \text{ feet}$$

The height of field $(H)$ is simply expressed as ¾ of the width. In this case:

$$H = 0.75 \times 5 = 3.75 \text{ feet}$$

Figure 3-10 gives field of view measurements taken at various lens-to-subject distances. It is important to note that this table is for use with vidicon camera lenses only.

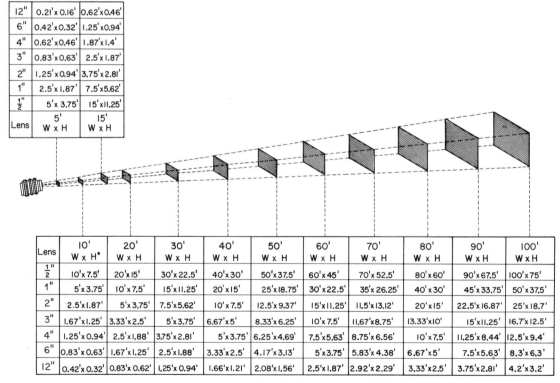

| Lens | 5' W x H | 15' W x H |
|---|---|---|
| 12" | 0.21'x 0.16' | 0.62'x0.46' |
| 6" | 0.42'x0.32' | 1.25'x0.94' |
| 4" | 0.62'x0.46' | 1.87'x1.4' |
| 3" | 0.83'x0.63' | 2.5'x1.87' |
| 2" | 1.25'x0.94' | 3.75'x2.81' |
| 1" | 2.5'x1.87' | 7.5'x5.62' |
| ½" | 5'x3.75' | 15'x11.25' |

| Lens | 10' W x H* | 20' W x H | 30' W x H | 40' W x H | 50' W x H | 60' W x H | 70' W x H | 80' W x H | 90' W x H | 100' W x H |
|---|---|---|---|---|---|---|---|---|---|---|
| ½" | 10'x7.5' | 20'x15' | 30'x22.5' | 40'x30' | 50'x37.5' | 60'x45' | 70'x52.5' | 80'x60' | 90'x67.5' | 100'x75' |
| 1" | 5'x3.75' | 10'x7.5' | 15'x11.25' | 20'x15' | 25'x18.75' | 30'x22.5' | 35'x26.25' | 40'x30' | 45'x33.75' | 50'x37.5' |
| 2" | 2.5'x1.87' | 5'x3.75' | 7.5'x5.62' | 10'x7.5' | 12.5'x9.37' | 15'x11.25' | 11.5'x13.12' | 20'x15' | 22.5'x16.87' | 25'x18.7' |
| 3" | 1.67'x1.25' | 3.33'x2.5' | 5'x3.75' | 6.67'x5' | 8.33'x6.25' | 10'x7.5' | 11.67'x8.75' | 13.33'x10' | 15'x11.25' | 16.7'x12.5' |
| 4" | 1.25'x0.94' | 2.5'x1.88' | 3.75'x2.81' | 5'x3.75' | 6.25'x4.69' | 7.5'x5.63' | 8.75'x6.56' | 10'x7.5' | 11.25'x8.44' | 12.5'x9.4' |
| 6" | 0.83'x0.63' | 1.67'x1.25' | 2.5'x1.88' | 3.33'x2.5' | 4.17'x3.13' | 5'x3.75' | 5.83'x4.38' | 6.67'x5' | 7.5'x5.63' | 8.3'x6.3' |
| 12" | 0.42'x0.32' | 0.83'x0.62' | 1.25'x0.94' | 1.66'x1.21' | 2.08'x1.56' | 2.5'x1.87' | 2.92'x2.29' | 3.33'x2.5' | 3.75'x2.81' | 4.2'x3.2' |

FIGURE 3-10. Field of view measurements of typical vidicon camera lenses at specific lens-to-object distances.

The formula just given also provides a convenient means of determining the lens focal length for a particular application. If the width of field is measured, and it is known how far the camera will be from the scene, the equation can be transposed as follows:

$$F = \frac{D}{2 \times W}$$

If the width of field is assumed to be 25 feet, and the lens-to-subject distance is 100 feet, then:

$$F = \frac{100 \text{ feet}}{2 \times 25 \text{ feet}} = 2\text{-inch focal length}$$

To assure accuracy, it is necessary that this measurement be made with the lens focused on an object that is quite distant from the lens. This will approximate a setting of infinity and satisfy the definition of focal length.

If scene width and lens focal length are known, the distance the camera must be placed from the scene is:

$$D = \frac{W \times F}{0.5}$$

## LENS SPEED

The *speed* of a lens is determined by the amount of light that it allows to pass through it. Therefore, lenses of large diameter are said to be faster than those whose diameter is smaller. Most television lenses have a variable opening, called an *iris,* located within the lens structure that can be used to control the amount of light transmitted to the focal plane. The function of the iris is the same as that of the iris within the human eye. By utilizing an iris the intensity of the image may be varied to compensate for varying conditions such as scene illumination, camera tube sensitivity, and the like.

By definition, *lens speed is a ratio between the lens opening and the focal length of the lens.* It is expressed as a number, or fractional number, referred to as the f/number. For example, an f/4.0 lens is one which has a maximum opening diameter that is ¼ the focal length. Simply restated, the focal length is four times the lens diameter. Figure 3-11 illustrates this relationship in a basic lens.

Lenses are generally marked in f/stop numbers imprinted on the iris ring of the lens body. The markings are arbitrarily chosen ratios of aperture to focal length, such as f/1.0, 1.4, 2.0, 2.8, 4.0, 5.6, 8, 11, 16, 22, etc. The smaller the f/stop number, the faster is the lens speed. The smaller numbers indicate a larger aperture, which makes more light available at the camera pickup tube.

When an iris is adjusted to its next numerically higher stop, the amount of light that is transmitted by the lens is typically reduced by one-half. For instance, if a setting of f/1.0 permits 200 foot-candles of light to fall on the vidicon faceplate, a setting of f/1.4 will decrease the light level to 100 foot-candles. Similarly, changing the setting to f/2.0 will reduce the level to 50 foot-candles, etc.

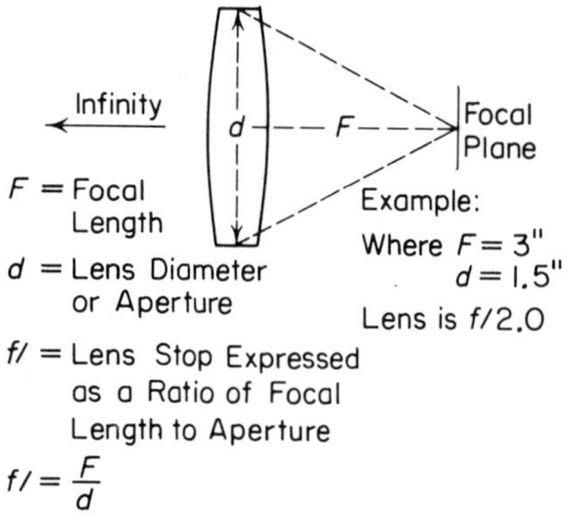

$F$ = Focal Length

$d$ = Lens Diameter or Aperture

$f/$ = Lens Stop Expressed as a Ratio of Focal Length to Aperture

$$f/ = \frac{F}{d}$$

Example:
Where $F = 3''$
$d = 1.5''$
Lens is $f/2.0$

FIGURE 3-11. Lens speed is a function of diameter and focal length.

At this point it is important to note that when a lens is advanced by two f/stops (from f/1.0 to f/1.4 to f/2.0), the amount of light transmitted is decreased by a factor of four. That is, the f/2.0 setting will pass only ¼ as much light as the f/1.0 setting.

A good example of the utility of the iris control is an outdoor scene being viewed by a television camera in bright sunlight. Let us assume that the lens is adjusted to a setting of f/11 to allow the proper amount of light to fall on the pickup tube. Should a cloud obscure the sun and reduce the scene illumination by one-half, reducing the lens iris one stop to a setting of f/8 would open the aperture sufficiently to offset the decrease in illumination.

## DEPTH OF FIELD

When a camera lens is focused on an object a specific distance away, there will also be other objects, both nearer and farther away, which will be in focus. The distance between the points nearest to and farthest from the camera, over which the scene appears to be in sharp focus, is called the *depth of field*.

Depth of field is dependent upon various factors. The most important of these are lens-to-subject distance, focal length, and aperture. Any variation in these factors will cause defocusing to some degree, but a very slight change will not be apparent to the human eye. It is simply a matter of determining when such variation has occurred to the extent that defocusing is evident to the human observer. Since the limits of depth of field depend upon an observer's definition of sharp focus, it is apparent that the limits are relative and subject to individual interpretation. When listing depth of field figures for lenses, it is necessary to rely on the visual acuity of an "average" observer.

The inability of the human eye to detect small changes in focus can be attributed to its internal construction. From a point about ten inches away, objects that are 0.01 inch in diameter or smaller all appear to be of equal size. Therefore, a point object observed at this distance may be defocused until the blurred image reaches about 0.01 inch in diameter without causing noticeable visual effects. This area, in which the image may be distorted but not noticeably so, is termed the *circle of confusion* for the human eye.

Television cameras also possess a circle of confusion characteristic, caused in large part by the granular structure of the photosensitive area and the cross-sectional area of the scanning beam. Coupling these characteristics of the vidicon and the human eye, plus limiting resolution capabilities of the lens and the remainder of the TV system, determines how much focus variation can occur unnoticed. This in turn determines what the depth of field will be.

To illustrate the effect of lens-to-subject distance on depth of field, Figure 3-12 shows the results of moving a point-source object toward the lens. Diameter of the defocused image increases as the object approaches. When the diameter of the image equals the limiting circle of confusion (shown in dotted lines) the object will be located at the near limit of the depth of field. Any further movement toward the lens causes *visible* defocusing. Similar results would be obtained by moving the object away from the point of optimum focus.

Figure 3-13 shows why the aperture setting of a lens affects depth of field. With the lens set to its widest opening, the defocused image occupies a large area which considerably exceeds the circle of confusion. If the aperture opening is reduced, those light rays which pass through the outer edges of the lens are cut off, narrowing the maximum convergence angle at the focal point. This reduces the area occupied by the defocused image until it is again confined within the limits of the circle of confusion, thereby demonstrating that a reduction in aperture opening increases the depth of field.

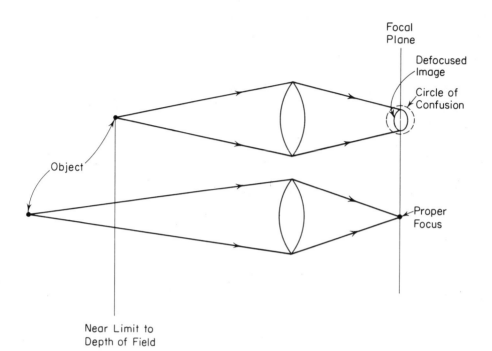

FIGURE 3-12. Circle of confusion determines depth of field in an optical system.

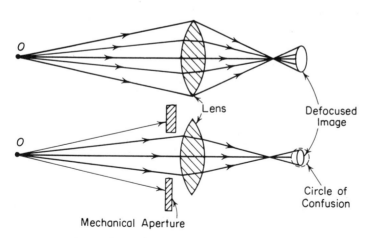

FIGURE 3-13. Depth of field is influenced by lens aperture setting.

Many lenses are calibrated to show the approximate depth of field with specific f/stop settings. Figure 3-14 shows a facsimile of a typical calibrated focus control ring on a vidicon camera lens. The arrow indicates the actual focus setting and the markings to each side of it indicate the depth of field limits when the lens iris is set to f/4.5 and f/22.

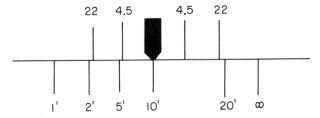

FIGURE 3-14. Lens focus element indicates depth of field for particular aperture settings.

Depth of field is also a function of the focal length of a lens. Wide-angle lenses exhibit a greater depth of field for a particular object distance than do their narrow-angle counterparts. This is true because of the greater shift in focal point that occurs in a narrow-angle lens with respect to a change in object distance. That is, the lens-to-subject distance becomes more critical as the focal length increases.

From the above it can be seen that depth of field is a rather complex parameter, dependent upon several variables, and calculations can become quite cumbersome. Most lens and television manufacturers publish depth of field figures that pertain to their products, making the determination of depth of field somewhat simpler. Table 3-2 shows the depth of field limits that may be expected from lenses that are widely used on vidicon television cameras. All of the parameters just discussed are taken into consideration in the table and their effects are quite apparent.

## COMPOUND LENSES

As has been suggested in the previous discussions, very little use is made of the simple, single-element lenses shown as illustrations. Although such a simple device does indeed form an image, its many drawbacks make it unsuitable for general use. To offset many of the drawbacks found in the single element lens, other lens elements are incorporated to correct or reduce objectionable effects.

In the discussion on refraction it was pointed out that a change in the density of the transmitting medium causes an angular change in the light path. It is an unfortunate fact, from a lens maker's point of view, that

**Table 3-2**

DEPTH OF FIELD LIMITS FOR
TYPICAL VIDICON CAMERA LENSES

**Object Distance**

| ½″ Lens | | | |
|---|---|---|---|
| | **f/1.5** | **f/5.6** | **f/16** |
| 5′ | 4′ to 7′ | 2′ to 60′ | 1.2′ to inf. |
| 10′ | 6′ to 35′ | 2.7′ to inf. | 1.5′ to inf. |
| 20′ | 8.5′ to inf. | 3′ to inf. | 1.7′ to inf. |
| 40′ | | 3.5′ to inf. | 1.8′ to inf. |

| 1″ Lens | | | |
|---|---|---|---|
| | **f/1.5** | **f/5.6** | **f/16** |
| 5′ | 4.6′ to 5.5′ | 3.8′ to 7.4′ | 2.6′ to inf. |
| 10′ | 8.5′ to 12.2′ | 6′ to 29.5′ | 3.4′ to inf. |
| 25′ | 17.3′ to 45.3′ | 9.4′ to inf. | 4.3′ to inf. |
| 50′ | 26.3′ to 490′ | 11.5′ to inf. | 4.7′ to inf. |

| 2″ Lens | | | |
|---|---|---|---|
| | **f/2.5** | **f/5.6** | **f/16** |
| 5′ | 4.8′ to 5.2′ | 4.6′ to 5.4′ | 4.1′ to 6.5′ |
| 10′ | 9.3′ to 10.8′ | 8.6′ to 12′ | 6.8′ to 18.9′ |
| 25′ | 21′ to 30.8′ | 17.6′ to 42.9′ | 11.4′ to inf. |
| 50′ | 36.3′ to 80.1′ | 27.2′ to 307′ | 14.7′ to inf. |

| 3″ Lens | | | |
|---|---|---|---|
| | **f/4** | **f/8** | **f/16** |
| 10′ | 9.5′ to 10.6′ | 9.1′ to 11.2′ | 8.3′ to 12.6′ |
| 20′ | 18.1′ to 22.3′ | 16.5′ to 25.3′ | 14.1′ to 34.5′ |
| 50′ | 39.6′ to 67.9′ | 33′ to 105′ | 24.3′ to inf. |
| inf. | 188′ to inf. | 93.8′ to inf. | 46.9′ to inf. |

| 4″ Lens | | | |
|---|---|---|---|
| | **f/4.5** | **f/8** | **f/16** |
| 20′ | 18.7′ to 21.4′ | 17.9′ to 22.7′ | 16.1′ to 26.3′ |
| 30′ | 27.2′ to 33.4′ | 25.4′ to 36.6′ | 22.1′ to 46.9′ |
| 50′ | 42.8′ to 60′ | 38.5′ to 71.2′ | 31.3′ to 124′ |
| inf. | 296′ to inf. | 166.7′ to inf. | 83.3′ to inf. |

| 6″ Lens | | | |
|---|---|---|---|
| | **f/4.5** | **f/8** | **f/16** |
| 20′ | 19.4′ to 20.6′ | 19′ to 21.1′ | 18.1′ to 22.4′ |
| 50′ | 46.5′ to 54′ | 44.1′ to 57.7′ | 39.5′ to 68.2′ |
| 100′ | 87′ to 118′ | 78.9′ to 136′ | 65.2′ to 214′ |
| inf. | 667′ to inf. | 375′ to inf. | 187.5′ to inf. |

different colors are not refracted to the same degree. The many colors viewed by a lens will each seek a somewhat different focal point, causing serious problems on the final image. This property of a simple lens runs closely parallel to the familiar action of a prism which deflects individual colors differently, and is termed *chromic aberration* when it occurs in a lens. Careful selection of the individual elements within a compound lens reduces the problem to a great extent. Such lenses are said to be *color corrected*.

Compound lenses also correct for a deficiency called *spherical aberration*. The phenomenon of spherical aberration arises from the fact that light rays that pass through a lens near its outer circumference tend to be refracted or bent too extensively, causing them to seek a shorter focal point. Spherical aberration becomes a greater problem as lens diameter increases and is accordingly harder to correct. This is one of the reasons for limiting the physical diameter of a lens. If spherical aberration is evident in a lens, it can be reduced considerably by setting the iris to a smaller aperture opening.

There are other aberrations found in simple lenses, such as astigmatism, coma, etc., and it falls to the compound lens to correct for these deficiencies. No lens yet produced can compensate perfectly for all of the possible errors, but modern design procedures have reduced their effects to a low level. If any defects are to be seen, it is generally true that a maximum opening of the iris will present them at their worst.

Another important function of a compound lens is to assure that the focal plane of lenses of different focal lengths all occur at precisely the same distance from the rear of the lens assembly. This feature is quite important to the camera designer since he must provide for a lens mount and position the camera tube so that a wide variety of lenses will be directly interchangeable. In vidicon cameras the distance between the camera tube faceplate and the rear lens element is normally 0.690 inch (Figure 3-15). This may seem a bit startling when considering a lens with a 0.5-inch focal length, since the vidicon faceplate is beyond the normal focal point, but elements are provided within the structure of the lens to extend the focal plane.

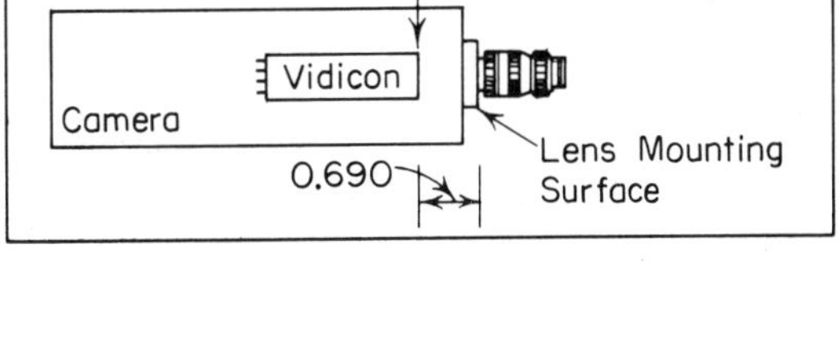

FIGURE 3-15. Proper lens-to-vidicon distance is important.

## LENS EXTENDERS

It is sometimes desirable to televise extreme closeup shots and to magnify small objects sufficiently to fill a TV monitor screen. Most lenses will not normally focus at the short distances necessary to accomplish this. They are able to focus down to a distance which is determined by the available forward movement of the front lens element. This movement is restricted on most lenses to the extent that the nearest limit of focus is at least several inches in front of the lens. Lenses of longer focal length, while providing a greater magnification capability, typically have minimum focus distances that are farther from the lens, and this makes them even more unsuitable for extreme closeup work. The use of extension tubes between lens and camera tube will allow focusing on objects at a closer range. (See Figure 3-16.)

Adding extension devices between lens and camera tube reduces the effective f/stop number. Depth of field is also drastically reduced. Because

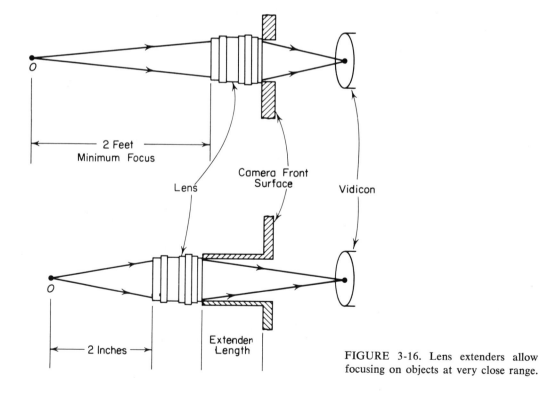

FIGURE 3-16. Lens extenders allow focusing on objects at very close range.

of this, supplementary lighting is often needed and the lens-to-subject distance must be held quite constant. When lens extenders of considerable length are used, depth of field is reduced to a mere fraction of an inch.

When it is desired to fill a monitor screen completely with an object of a particular size, there is considerable latitude of choice between lens focal length and the particular extender length that might be used. Table 3-3 gives examples of typical object widths, in order of increasing size, which might be viewed over the entire width of a monitor display, together with a lens focal length and extender length which would accomplish the desired magnification. The table assumes that the lenses are set to their minimum focusing distance and describes the lens-to-subject distance necessary for the particular combination.

For the more mathematically inclined, it is possible to calculate the length of extension needed for a particular application in the following manner:

**Table 3-3**

TYPICAL EXTENDER-LENS COMBINATIONS

| Object Width | Lens to Subject Distance | Lens Focal Length | Extender Length |
|---|---|---|---|
| ¼″ | 3⅛″ | 3″ | 6″ |
| ⁵⁄₁₆″ | 9¹⁄₁₆″ | 4″ | 6″ |
| ⅜″ | 9⁹⁄₁₆″ | 4″ | 5″ |
| ⅜″ | 3⅝″ | 3″ | 4″ |
| ⁷⁄₁₆″ | 1⅞″ | 2″ | 2″ |
| ⁷⁄₁₆″ | 14⅝″ | 6″ | 6″ |
| ½″ | 4⅜″ | 3″ | 3″ |
| ½″ | 10³⁄₁₆″ | 4″ | 4″ |
| ⁹⁄₁₆″ | 15⅝″ | 6″ | 5″ |
| ⅝″ | 11⁵⁄₁₆″ | 4″ | 3″ |
| ¾″ | 12⅛″ | 4″ | 2½″ |
| ¾″ | 17⅛″ | 6″ | 4″ |
| ⅞″ | 3⅜″ | 2″ | 1″ |
| ⅞″ | 13¼″ | 4″ | 2″ |
| 1″ | 19⅝″ | 6″ | 3″ |
| 1⅛″ | 15″ | 4″ | 1½″ |
| 1³⁄₁₆″ | 8⅝″ | 3″ | 1″ |
| 1⅜″ | 24″ | 6″ | 2″ |
| 1½″ | 5⅝″ | 2″ | ½″ |
| 1⁹⁄₁₆″ | 18⁵⁄₁₆″ | 4″ | 1″ |
| 2″ | 13⅛″ | 3″ | ½″ |
| 2⅜″ | 36″ | 6″ | 1″ |
| 2¹⁵⁄₃₂″ | 25″ | 4″ | ½″ |

### Length of Extension Equation

$$d = \frac{f^2}{U - f}$$

where

$d = $ Extension

$f = $ Focal length

$U = $ Distance from lens to object

EXAMPLE:

> 50 mm lens
>
> 19½ inches from object
>
> (Convert to centimeters)

*Extension must be:*

$$\frac{25}{50 - 5} = \frac{25}{45} = 0.56 \text{ centimeter or } 0.22 \text{ inch}$$

Note: Focusing adjustment compensates to some degree.

## ZOOM LENSES

A *zoom lens* (Figure 3-17) is a lens that is variable in focal length over a wide range. Probably everyone who has attended movies or watched home television has become familiar with the effects of a zoom lens. Indeed, the lens has become so popular that many home movie cameras are equipped with them as permanent fixtures.

Figure 3-18 illustrates the field-of-view capabilities of a particular 10:1 zoom lens. The 10:1 designation indicates the variation of focal length that can be accomplished. In the diagram, the object distance is set at 20 feet. With the lens zoomed to its narrowest angle of 4.3 degrees, the field of view is 1.6 feet. If the zoom element of the lens is moved until the widest angle of 43 degrees is obtained, the field of view becomes 16 feet.

The variable focal length of a zoom lens is achieved by moving individual lens elements within the assembly. Figure 3-19 illustrates a typical zoom lens with four separate compound components, each containing several lens elements. Component A is movable over a small range and is used as the focus control. Component B is the movable element that varies focal length and results in the zoom effect. During the movement of B, C also

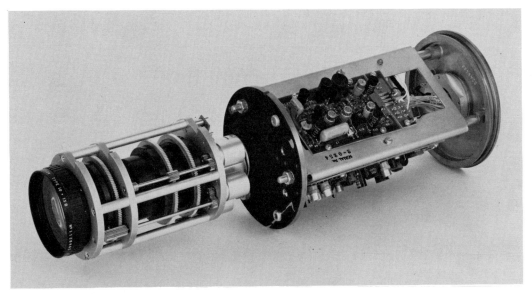

FIGURE 3-17. A 10:1 zoom lens mounted on a vidicon camera. This one is motorized for remote operation.

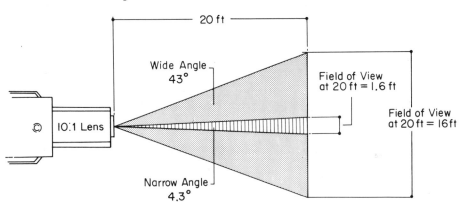

FIGURE 3-18. Field of view capabilities of a typical 10:1 zoom lens.

moves automatically to keep the focal plane stationary and thereby elimi-nate any defocusing during the zoom process. Component D is a stationary element within the assembly.

For a zoom lens to track properly (remain in focus throughout the entire zoom range) it is imperative that the lens be critically positioned

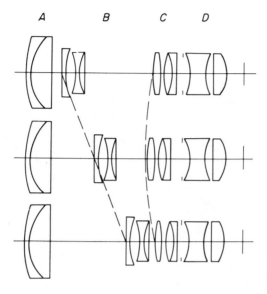

FIGURE 3-19. A simplified diagram of the elements of a zoom lens.

with respect to the faceplate of the camera pickup tube. That is, for each particular zoom lens there is a distance between the rear lens element and the pickup tube that must be accurately maintained. It is possible, by use of the focus control, to obtain proper focus for any particular zoom setting without achieving this critical distance, but in such instances causing the lens to zoom will result in defocusing of the image.

To *track* a zoom lens simply means to position it with respect to the camera tube so that proper focus is maintained during the zoom process. To achieve this, it is usually easiest if procedures similar to the following are used:

1. Open the iris to its widest aperture. This will reduce the depth of field for a more critical adjustment.
2. Position the zoom control to one end of its travel (at its maximum wide-angle position, for instance).
3. Direct the camera at an object that is a considerable distance from the camera (the farther, the better). Set the focus control to the infinity position (for all practical purposes, a far-distant object is at infinity).
4. Observing the televised scene on a monitor, simply position the lens by screwing it into the lens mount until sharp focus is achieved. The lens should now remain in focus throughout its entire zoom range.

Zoom lenses can theoretically simulate any fixed lens that has a focal length that lies within its zoom range. This, plus the dramatic effects it can achieve, make it an extremely popular device. However, it should not be considered the answer to all applications. Zoom lenses are not fast lenses. In areas where illumination is poor, it may be inadequate to the demand, and the user might be better advised to use faster fixed-focal length lenses mounted on a lens turret (Figure 3-20). Zoom lenses are also subject to a greater severity of aberrations because of the difficulty in compensating for them with the number of movable optical elements involved.

COURTESY ALBION OPTICAL COMPANY.

FIGURE 3-20. Four lenses as they might be located on a turret.

## SPECIALIZED OPTICS

There are many applications which require optics that are not generally available as production items. Many times it is necessary to tailor the optics to the specific application and, while this may be an expensive proposition, it is often the only alternative. The possible configurations that such optics may take are almost limitless and are probably best described by way of example:

Those who use underwater television cameras have long been disturbed by the apparent magnification that is caused when the camera is submerged. This effect exists because of the change in index of refraction

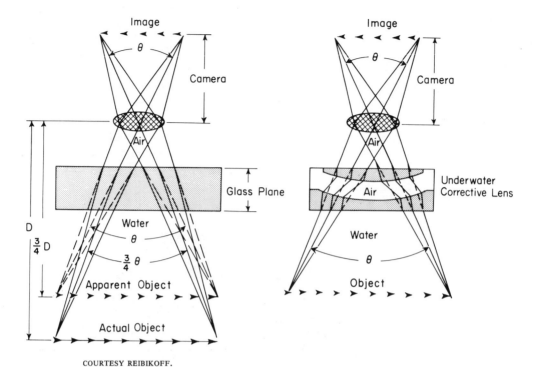

FIGURE 3-21. Optical correction for underwater cameras eliminates distortion caused by the different index of refraction possessed by the water.

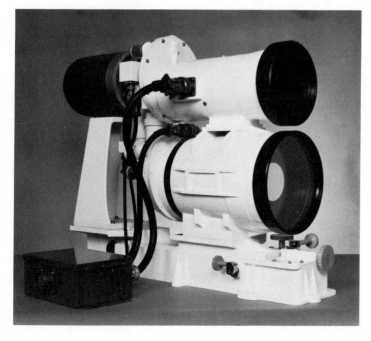

FIGURE 3-22. A radar boresighting system. TV camera is the dark cylinder at top left.

(water/glass vs. air/glass) in the submerged environment. One means of correcting for this distortion is to use a lens system that *optically* causes the water to appear as if it were air (Figure 3-21). In this case the corrective lens also forms the front viewing port of a camera housing and must therefore be able to withstand the considerable pressures encountered at depth.

Another example of specialized optics is shown in Figure 3-22. This is a lens designed for use with a radar bore-sighting system, where extreme focal length and accuracy of positioning are imperative. It consists of two lenses—an 80-inch focal length Reflectar and a 3.5- to 35-inch focal length zoom lens. The television camera is the dark cylinder mounted at the top rear of the optical system.

Lens requirements such as these place demands upon lens manufacturers that result in advances in the "state of the art" and contribute to improvements in standard lenses as well.

# 4

# CAMERA TUBES

The television camera tube may be thought of as the eye within the body of a television system. For such an analogy to be correct, the tube must possess characteristics that are similar to its human counterpart. Some of its more important functions must be: a sensitivity to visible light, a wide dynamic range with respect to light intensity, and an ability to resolve detail when viewing a multielement scene.

Present day camera tubes fulfill the above requirements in a worthy manner. In some areas they far exceed the capabilities of the human eye. For instance, there are tubes in use that are so sensitive that they present useful output under conditions that the eye interprets as complete darkness. This represents a considerable advancement in the "state of the art" over early camera tubes that required 1000 foot-candles of illumination to produce a usable picture.

Several types of photosensitive camera tubes have been developed to meet the needs of various applications. These tubes may vary widely in physical size and operating characteristics. The difference between tubes of different type lies primarily in the composition of the photosensitive material upon which the light falls, and the means used to extract the electrical information that is thereby produced.

The two major types of camera pickup tubes that dominated the television industry until the mid-1960's were the *vidicon* and the *image-orthicon*. Of these two, the vidicon still accounts for the vast majority of television camera tubes in use. The image-orthicon finds its major use in broadcast applications, while the vidicon has been largely responsible for the growth of the closed circuit television industry. Other tubes, similar to the vidicon, but using different types of photosensitive material on the tube faceplate, have recently appeared and are quite popular. These newer tubes possess definite advantages in many applications, and the number and type of devices will undoubtedly increase rapidly to meet the ever widening demands of those who employ television systems.

Generally speaking, an understanding of the vidicon and image-orthicon provides the basic mechanics utilized in the operation of nearly all types of television camera tubes. In the following discussion primary emphasis is placed on the vidicon because of its predominant use, especially in the field of closed circuit television.

## THE IMAGE-ORTHICON

The image-orthicon (Figure 4-1) finds its primary use in broadcast studios. Certain industrial and military requirements also find its capabilities desirable, even though the size and power requirements of image-orthicon cameras are generally greater than that of similar vidicon equipment.

Broadcasters especially appreciate the halo effect of the image-orthicon. This is probably best described as a black border or edge that is produced around the outlines of a bright object. While it is not a true representation of the viewed object, it does tend to enhance the image by making it appear to be better defined or sharper in appearance. Because the human eye also produces a similar effect when viewing bright objects, the brain is conditioned to accept the phenomenon as natural.

Figure 4-2 is a cross-sectional view of an image-orthicon tube. It is shown as being composed of three basic sections—an image section, a scanning section, and a multiplier section.

### The Image Section

The image section is made up of three main elements: the photocathode, an accelerator, and the target. The photocathode is the light-sensitive element that constitutes the faceplate of the tube. In image-orthicon tubes, it is generally a semitransparent material that is deposited on the rear side of the front glass. Light from a scene is focused upon the photocathode which then emits electrons from each illuminated area in proportion to the intensity of the light.

The electrons that are emitted from the photocathode are focused onto

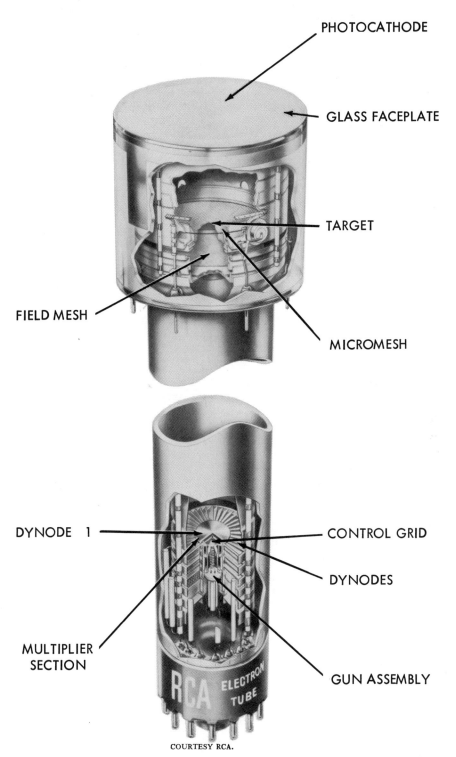

PHOTOCATHODE

GLASS FACEPLATE

TARGET

FIELD MESH

MICROMESH

DYNODE 1

CONTROL GRID

DYNODES

MULTIPLIER
SECTION

GUN ASSEMBLY

FIGURE 4-1. An exploded view of an image-orthicon.

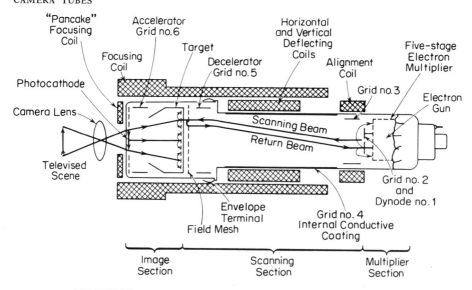

FIGURE 4-2. Image-orthicon cross-section.

the target by means of a magnetic field produced by an external coil, and the combined effects of proper accelerator and photocathode voltages. Thus, the original light image is converted into a multielement electron field that is directed at the target.

The target consists of a very thin glass disc which has a fine mesh screen positioned close to its photocathode side. When electrons from the photocathode strike the target, their impact causes the target to emit additional electrons. This action is termed *secondary emission,* and the number of secondary electrons emitted is generally greater than, but proportional to, the density of the original electron bombardment. The secondary electrons are collected by the adjacent mesh screen and drained off to a power supply. The target is thus left with a positive static charge pattern, which corresponds to the original light image on the photocathode. Because the glass that constitutes the target is very thin, a similar charge pattern is also produced on the rear side of the glass.

**The Scanning Section**

The reverse side of the target glass is scanned by a low velocity electron beam produced by an electron gun in the scanning section. This gun contains a thermionic cathode that emits electrons, a control grid (grid No. 1) to control the density of the electron beam, and an accelerating grid (grid No. 2). The beam is focused at the target by a magnetic field from an external focusing coil and by the action of the electrostatic field of grid No. 4. Deflection of the beam for scanning is accomplished magnetically by use of externally mounted deflection coils (see Chapter Five).

By proper adjustment of potentials, the beam is caused to approach the target perpendicularly and with essentially zero velocity. The electrons stop their forward motion at the surface of the target glass and are turned back to be focused into the multiplier section, except when they approach positive charge sections of the pattern impressed on the glass. When this happens, electrons are taken from the beam to neutralize the positive charge that exists. Thus, the beam that returns to the multiplier section varies in density in proportion to the charged areas across which it is scanned. The return beam is therefore modulated with video information obtained from the target area.

### The Multiplier Section

The multiplier section also uses the principles of secondary emission in its operation. The return beam is caused to strike dynode No. 1, resulting in the release of many other electrons. These secondary electrons are directed to dynode No. 2, where the process is repeated, releasing even more electrons (dynodes 2 through 5 are not shown in Figure 4-2, but are schematically illustrated in Figure 4-3). Similar action takes place at dynodes 3 through 5, resulting in an effective gain in the number of electrons by a factor of 500 or so. The stream of electrons emitted from dynode No. 5 are attracted to the anode, which serves as the output terminal of the image-orthicon.

### Characteristics

Figure 4-4 shows the spectral response curve of a typical image-orthicon. It can be seen that its major sensitivity lies in the violet and blue regions of the visible spectrum. It is also sensitive to radiations in the ultraviolet region, although the poor ultraviolet transmission characteristics of lenses keep this from being much of a problem. It shows its least response to red.

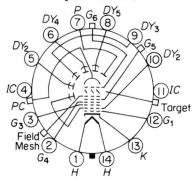

Basing Diagram

Bottom View

Direction of Light:
Perpendicular to
Large End of Tube

Small-shell Diheptal 14-pin Base

Pin 1: Heater
Pin 2: Grid No.4
Pin 3: Grid No. 3
Pin 4: Internal Connec-
tion — do not use
Pin 5: Dynode No. 2
Pin 6: Dynode No. 4
Pin 7: Anode
Pin 8: Dynode No. 5

Pin 9: Dynode No. 3
Pin 10: Dynode No.1,
Grid No. 2
Pin 11: Internal Connec-
tion — do not use
Pin 12: Grid No. 1
Pin 13: Cathode

Envelope Terminals

Terminal Over Pin 2: Field Mesh
Terminal Over Pin 4: Photocathode
Terminal on Side
of Envelope
Opposite Base Key: Grid No. 6
Terminal Over Pin 9: Grid No. 5
Terminal Over Pin 11: Target

FIGURE 4-3. Image-orthicon sche-matic.

Figure 4-5 illustrates the *light transfer characteristics* of the image-orthicon, and demonstrates the sensitivity of the device. The light transfer characteristic is a measure of output signal current with respect to illumination. The graph might appear linear at first glance, but it should be noted that it is plotted on log paper, so it should be understood that the signal output is not a linear function of the light input. As a matter of fact, circuitry is often included in camera systems to compensate for this form of amplitude distortion. These circuits, called "gamma correction" circuits

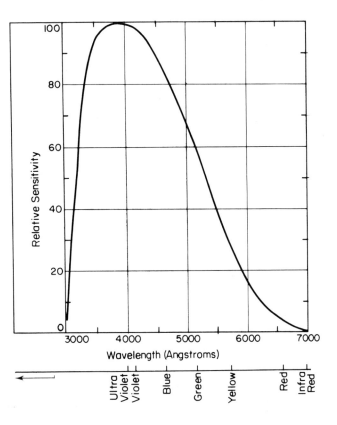

FIGURE 4-4. Spectral response curve
of RCA 8674 image-orthicon.

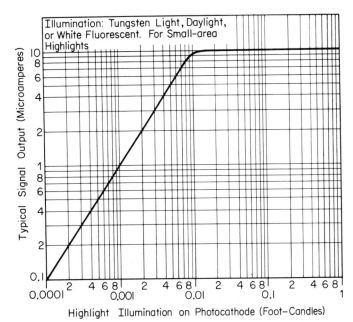

FIGURE 4-5. Light transfer characteristics of RCA 8674 image-orthicon.

69

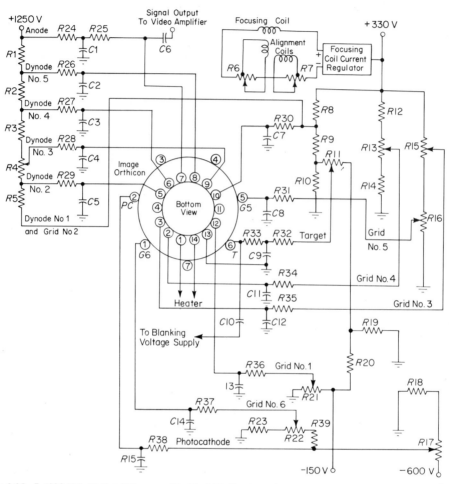

C1, C2: 0.05 μF, 1600–Volt Working Voltage

C3: 0.01 μF Mica, 1600–Volt Working Voltage

C4: 0.001 μF Mica, 1000–Volt Working Voltage

C5: 0.001 μF Mica, 600–Volt Working Voltage

C6: 0.03 μF, 1600–Volt Working Voltage
  See Text Under Voltage Divider

C7, C8, C9, C11, C12, C13, C14, C15: 0.001 μF
  Mica, 400–Volt Working Voltage

C10: 1 μF, 400–Volt Working Voltage

R1: 47,000 Ohms, $\frac{1}{2}$ Watt

R2, R3: 220,000 Ohms, $\frac{1}{2}$ Watt

R4: 220,000–Ohm Potentiometer, 1 Watt

R5: 270,000 Ohms, $\frac{1}{2}$ Watt

R6, R7: 200–Ohm Potentiometer,
  Center–tapped, 2 Watts

R8: 5100 Ohms, $\frac{1}{2}$ Watt

R9: 51,000 Ohms, 2 Watts

R10: 510 Ohms, $\frac{1}{2}$ Watt

R11: 150,000–Ohm Potentiometer, $\frac{1}{2}$ Watt

R12: 125,000 Ohms, $\frac{1}{2}$ Watt

R13: 75,000–Ohm Potentiometer, 1 Watt

R14: 125,000 Ohms, $\frac{1}{2}$ Watt

R15: 250,000–Ohm Potentiometer, 1 Watt

R16: 200,000–Ohm Potentiometer, 1 Watt

R17: 250,000–Ohm Potentiometer, 1 Watt

R18: 470,000 Ohms, 1 Watt

R19: 11,000 Ohms, $\frac{1}{2}$ Watt

R20: 310,000 Ohms, $\frac{1}{2}$ Watt

R21: 500,000–Ohm Potentiometer, 1 Watt

R22: 150,000–Ohm Potentiometer, 1 Watt

R23: 220,000 Ohms, $\frac{1}{2}$ Watt

R24: 47,000 Ohms, $\frac{1}{2}$ Watt

R25: 20,000 Ohms, $\frac{1}{2}$ Watt

R26, R27: 100,000 Ohms, $\frac{1}{2}$ Watt

R28, R29, R31, R32, R34, R35, R36,
  R37, R38: 200,000 Ohms, $\frac{1}{2}$ Watt

R30: 50,000 Ohms, $\frac{1}{2}$ Watt

R33: 100,000 Ohms, $\frac{1}{2}$ Watt

R39: 270,000 Ohms, $\frac{1}{2}$ Watt

FIGURE 4-6. Typical control circuitry for an image-orthicon.

compensate by introducing nonlinearity of amplitude response that is just the opposite of that of the image-orthicon, and the effect is essentially cancelled in the output.

Image-orthicon sensitivity is such that, with proper illumination, it can have an equivalent ASA exposure index of well in excess of 5000, depending upon the particular type that is used (compared at a shutter speed of $\frac{1}{60}$ second, which is the field rate of the television system). The tube is also somewhat analogous to film in that it is a fixed-sensitivity device. That is, to compensate for changes in light level at the scene it is necessary to reduce the light transmission to the tube *optically* by use of filters or by varying the iris element of the lens. (Vidicons, on the other hand, can compensate electrically.)

The knee of the curve shown in Figure 4-5 is reached when the light level causes the target to become fully charged with respect to the screen (or mesh) between successive scans of the electron beam. However, it is when the tube is operated slightly above this knee that the desirable halo effect (a black border surrounding bright objects) occurs. The phenomenon occurs because electrons emitted from the saturated target fall back on the target, producing an area of reduced potential around the perimeter of highlighted sections of the target. This results in a dark edge, or border, around brighter subjects.

The schematic shown in Figure 4-6 is a typical voltage divider arrangement for providing the proper potentials to an image-orthicon. It should be noted that the voltage applied to dynode No. 3 is made variable by the action of potentiometer $R4$. This controls the amount of secondary emission that will take place on this dynode and forms an effective gain control for the tube. It is apparent from the number of controls involved that the setup and initial operating procedures for the image-orthicon are somewhat complicated. This, coupled with its power requirements and relative size (2 to 4½ inches in diameter), have generally limited its use outside broadcast and certain specialized applications.

## THE VIDICON

The vidicon came into general use in the early 1950's and gained immediate popularity because of its small size (1-inch diameter) and ease of operation. It is also less expensive than the image-orthicon. The vidicon has enjoyed great success in the closed circuit television field and advances in vidicon technology have also made it attractive for use in many broadcast applications.

### Principles of Operation

Figure 4-7 illustrates the structural configuration of a typical vidicon. It is simpler in operation than the image-orthicon and does not rely upon

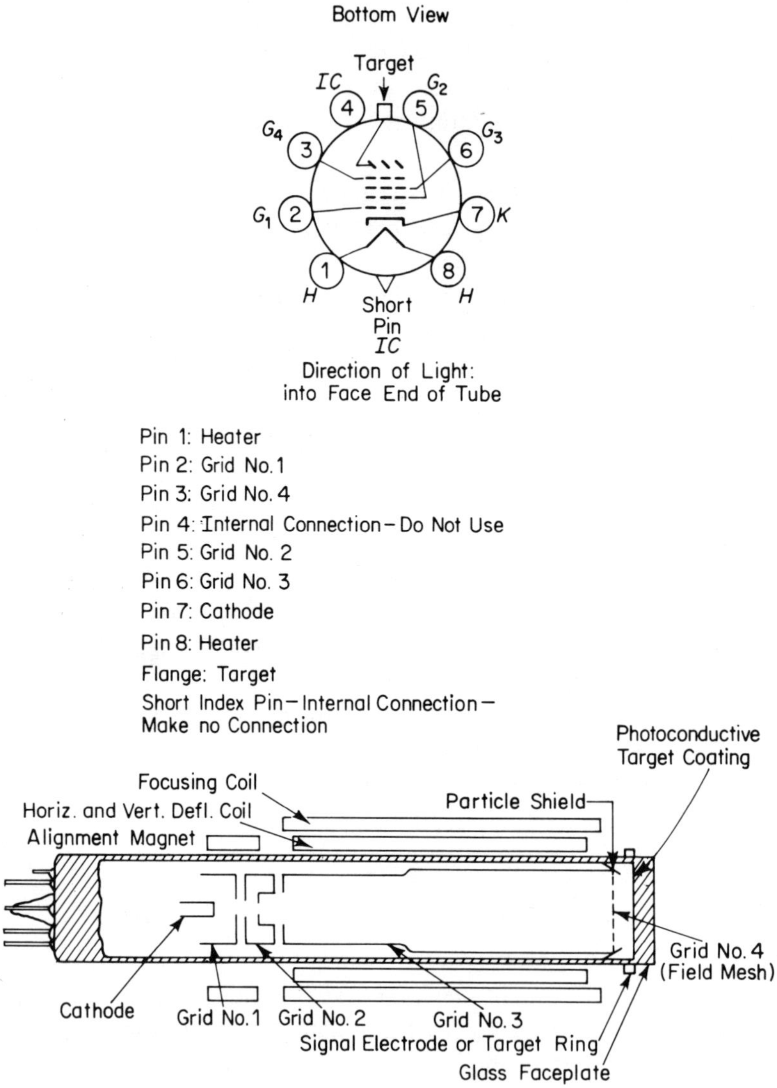
Bottom View
Target
IC
$G_2$
4
5
$G_4$
3
6
$G_3$
$G_1$
2
7
K
1
8
H
H
Short
Pin
IC
Direction of Light:
into Face End of Tube
Pin 1: Heater
Pin 2: Grid No. 1
Pin 3: Grid No. 4
Pin 4: Internal Connection – Do Not Use
Pin 5: Grid No. 2
Pin 6: Grid No. 3
Pin 7: Cathode
Pin 8: Heater
Flange: Target
Short Index Pin – Internal Connection – Make no Connection
Photoconductive Target Coating
Focusing Coil
Horiz. and Vert. Defl. Coil
Alignment Magnet
Particle Shield
Grid No. 4 (Field Mesh)
Cathode
Grid No. 1
Grid No. 2
Grid No. 3
Signal Electrode or Target Ring
Glass Faceplate

COURTESY RCA.

FIGURE 4-7. Cross-section of a vidicon tube.

secondary emission to form the output signal. With the exception of the scanning function, all of the processes involved in generating the video signal occur at the front surface of the tube.

The photosensitive element, or target of the vidicon, consists of a transparent conducting film (signal electrode) placed on the inner surface of the faceplate glass and a thin photoconductive layer deposited on the conductive film. The photoconductive layer can be thought of as a large number of individual target elements, each consisting of a capacitor paralleled with a light-dependent resistor (Figure 4-8). One end of each of these

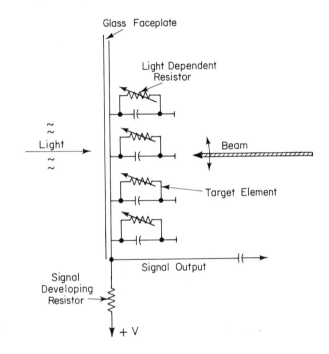

FIGURE 4-8. Schematic representation of vidicon target area.

target elements is connected to the signal electrode and the other end is unterminated, facing the beam.

With no light directed at the target, the parallel resistors maintain a high value of resistance. When light is present their resistance drops to a value that is inversely proportional to the intensity of the light. When a positive voltage is applied to the signal electrode, all of the capacitors charge toward the potential of the applied voltage. Of course, each time the beam scans across the target elements they are discharged by the deposited electrons. If the target is left dark, the capacitors will have little time to charge, due to the high parallel resistance and the rapid scanning cycle ($\frac{1}{30}$ of a second). However, if some of the target elements are illuminated, the lower resistance will allow faster charge and the capacitor plates which face the beam will assume a positive potential. Because of this action, a charge pattern is established over the reverse side of the target area. The pattern formed is proportional in intensity to the amount of light impressed on the front side of the target.

The beam that scans the target consists of low velocity electrons that are deposited on the positively charged areas, returning them to the negative potential of the beam. This causes a capacitive current to be generated that flows through the signal electrode to an external signal-developing resistor. The magnitude of the current developed is proportional to the rear-surface charge of the element being scanned. Since this surface charge is a function of the amount of light falling on the target elements, the output is an electrical representation of the optical image.

The beam is generated by a thermionic cathode and its density is regulated by a control grid (grid No. 1). Beam acceleration is achieved by the action of grid No. 2. It is focused at the surface of the target by the combined action of a uniform magnetic field of an external focusing coil and the electrostatic field of grid No. 3. Grid No. 4 serves to provide a uniform decelerating field between itself and the photoconductive layer so that the electron beam will tend to approach in a direction perpendicular to the target. The low velocity of the beam at the target is due to the low voltages (generally below 30 volts) used for the signal electrode.

Beam deflection is usually accomplished by means of magnetic fields generated by external deflection coils. Vidicons are available that utilize electrostatic deflection, but magnetically deflected tubes are more popular because of their better resolution capabilities and economy.

### Vidicon Characteristics

An important feature of the vidicon is its ability to operate at different levels of sensitivity. This is accomplished by varying the target voltage with respect to the highlight illumination on the faceplate. Thus, it is not necessary to place reliance solely on the optics to compensate for changes in scene illumination.

Using the RCA 8507A vidicon as an example, a target voltage of 10 to 22 volts will provide for proper operation with a highlight illumination of 10 foot-candles on the faceplate. When the scene is changed to one where the light illumination is only 0.1 foot-candle, increasing target voltage to approximately 30 to 60 volts will increase the sensitivity to the degree necessary to compensate for the change in light level. This ability, when coupled with an optical system that has a variable iris, gives the vidicon camera an extremely wide operating range with respect to changing levels of illumination.

Figure 4-9 shows typical voltage divider networks that may be used to obtain the proper vidicon operating potentials. Three variable controls are provided for adjustments to beam, target, and electrostatic focus elements. As previously stated, the target voltage is generally adjusted to suit lighting conditions. Adjustment of beam intensity is governed by the value of the applied target voltage and the scene highlights. The beam must have adequate intensity to discharge the highlight elements of the photoconductive surface to cathode potential on each scan. When insufficient beam is provided, only the lowlight elements of the photoconductor will be returned to cathode potential, and the highlight elements will not be sufficiently discharged to provide their proper amount of output signal. Consequently, the picture highlights will all appear to have the same relative amplitude and a washed-out picture with little detail in the highlight areas will result. In this condition, the vidicon is said to be "beam starved."

Insufficient beam current will also allow the individual target elements, which normally charge to only a small portion of the applied target voltage, to increase their charge gradually to nearly full target potential. This occurs because the highlighted target areas are not returned to cathode potential during each scan and they resume charging at a somewhat higher potential after each passage of the beam. When the highlight illumination is removed, it takes several passages of the beam to completely discharge those areas affected. As a result, the remaining charge causes an image to be produced for a period of time after the illumination is removed and the high-brightness areas tend to stick. When viewing a bright moving object, this condition causes a white tailing or smearing effect to be observed at its trailing edge.

Although it is important to provide sufficient beam current, it should be noted that too much beam intensity will increase the size of the scanning spot (that area where the beam strikes the target), and a loss of resolution will result.

The electrostatic focus control (grids 3 & 4) allows focusing of the beam to a very narrow point as it approaches the target surface. It is generally used in conjunction with an externally generated magnetic field to

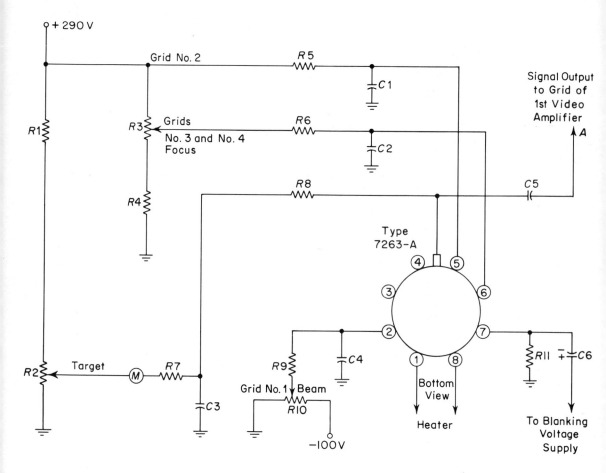

C1 C4: 0.1 μF, 300 Volts (Working Voltage)

C2: 0.1 μF, 300 Volts (Working Voltage)

C3 C5: 0.1 μF, 200 Volts (Working Voltage)

C6: 4 μF, Electrolytic, 300 Volts (Working Voltage)

M: Target–Current Meter

R1: 120,000 Ohms, $\frac{1}{2}$ Watt

R2: 100,000-Ohm Potentiometer, 2 Watts

R3: 50,000-Ohm Potentiometer, 2 Watts

R4: 70,000 Ohms, $\frac{1}{2}$ Watt

R5: 5000 Ohms, $\frac{1}{2}$ Watt

R6: 10,000 Ohms, $\frac{1}{2}$ Watt

R7: 200,000 Ohms, $\frac{1}{2}$ Watt

R8: 50,000 Ohms, $\frac{1}{2}$ Watt, Noninductive

R9: 100,000 Ohms, $\frac{1}{2}$ Watt

R10: 500,000-Ohm Potentiometer, 2 Watts

R11: 1000 Ohms, $\frac{1}{2}$ Watt, Noninductive

FIGURE 4-9. Control circuitry for a typical vidicon.

achieve proper spot size. Varying either the magnetic or electrostatic fields will change the focus. Once set, the magnetic focus is generally not changed and any variations in focus are corrected by adjusting the grid potentials to compensate.

The vidicon target material can be permanently damaged by exposure to the focused image of extremely bright objects. The time that is required for such damage to occur becomes shorter as the object brightness is increased. If a camera is directed at the sun, for example, the focused image of that intense source will cause immediate and permanent damage, *even if the camera is not energized at the time*. Objects of lower intensity can also cause "burns" on the target material if the vidicon is exposed to them for a sufficient length of time. It is always a good practice to cap the lens of a camera when it is not in use, especially if the camera is located outdoors or in a high brightness area.

Vidicon resolution capabilities may vary over a wide range, depending upon the applications for which each tube is designed. Some tubes can display well over 1000 lines of resolution. However, as the resolution elements become smaller, the signal output from the tube decreases in amplitude. The primary cause of this is the finite size of the beam spot on the photoconductor. When resolution elements become smaller than the cross-section of the beam, the beam not only discharges the element of resolution, but some of the surrounding material as well. The output of this minute area is therefore the sum of currents from all of the target elements that occur within the cross-sectional area of the beam. Thus, small bits of resolution are integrated with a portion of the background and the result is a signal whose amplitude has been compromised.

Figure 4-10 shows a graph that illustrates the relative amplitude response of an 8507A vidicon with respect to the number of resolution elements (TV line number) being viewed. Because of this vidicon characteristic, compensation is often included in video amplifiers of television cameras to offset the decreased response at the higher frequencies.

Vidicon *dark current* is an important, and generally undesirable, feature of the vidicon tube. It is defined as that current generated by scanning the target area of a tube whose face is in complete darkness. The current is generated because the photoconductor is not a perfect insulator when darkened (although that would be an ideal situation), and tends to build up a small amount of positive charge at its rear surface between successive scans by the beam. Dark current increases with an increase in target volt-

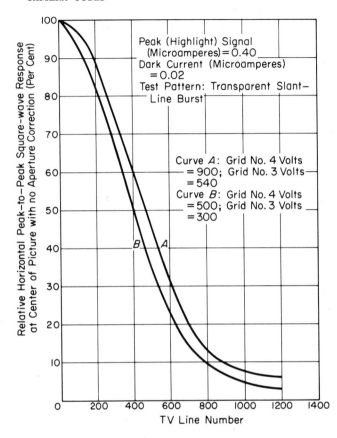

Relative Horizontal Peak-to-Peak Square-wave Response at Center of Picture with no Aperture Correction (Per Cent)

Peak (Highlight) Signal
(Microamperes) = 0.40
Dark Current (Microamperes)
= 0.02
Test Pattern: Transparent Slant–
Line Burst

Curve A: Grid No. 4 Volts
= 900; Grid No. 3 Volts
= 540
Curve B: Grid No. 4 Volts
= 500; Grid No. 3 Volts
= 300

TV Line Number

COURTESY RCA.

FIGURE 4-10. Relative amplitude response of RCA 8507 Vidicon.

age, since the higher potential forces a faster charging rate for the target elements.

Vidicon sensitivity is usually specified in relation to dark current. This is done because the dark current offers a tangible foundation upon which to base such a parameter. For instance, if several tubes were set up to generate the same amount of dark current and then exposed to identical light sources, it would be a simple matter to compare the signal outputs of the tubes. Obviously, those with greater output amplitude would possess more sensitivity than those whose output amplitude was lower.

The light transfer characteristic of the vidicon differs considerably from the image-orthicon previously shown. In addition, the characteristic curve varies with differences in dark current, as shown in Figure 4-11. Again, it should be pointed out that the curve is by no means linear but follows a *log* function. The average *gamma,* or slope, of the light transfer

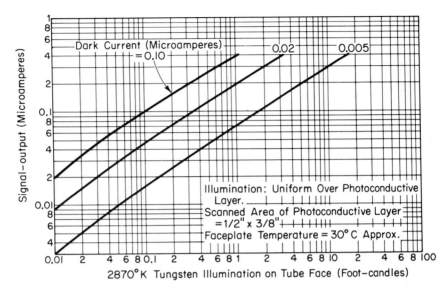

FIGURE 4-11. Light transfer characteristics of a typical vidicon.

characteristic is approximately 0.65. It is interesting to note on the diagram that the slope remains relatively constant over rather wide variations in dark current.

It is important to note that the gamma of a vidicon tube is approximately the complement of the transfer characteristic of a monochrome or color receiver picture tube. The two produce a picture having proper tone rendition, and no "gamma correcting" circuitry is needed. That is, the nonlinearity of light transfer in the vidicon is just the opposite of picture tube nonlinearity, and the two cancel.

An additional parameter that is generally related to dark current is that of *vidicon lag*. The lag, or persistence, is a measure of the decay in signal-output current after the illumination source is removed. Figure 4-12 shows the lag characteristic of a particular vidicon for two different values of dark current.

The *spectral response* of vidicons is generally quite close to that of the human eye. Figure 4-13 illustrates a response curve that typifies the spectral characteristics of many popular vidicons. By proper choice of target material, vidicons can be produced that are sensitive to the infrared or ultraviolet regions, making them useful in a variety of scientific, industrial, and military applications.

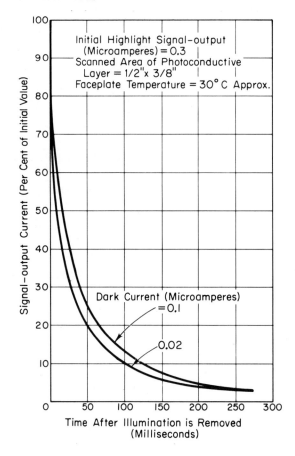

Initial Highlight Signal-output
(Microamperes) = 0.3
Scanned Area of Photoconductive
Layer = 1/2"x 3/8"
Faceplate Temperature = 30° C Approx.

Dark Current (Microamperes)
= 0.1

0.02

Time After Illumination is Removed
(Milliseconds)

Signal-output Current (Per Cent of Initial Value)

COURTESY RCA.

FIGURE 4-12. Vidicon persistence characteristics.

## RECENT CAMERA TUBES

It was previously mentioned that newer types of camera tubes, similar to the vidicon, but differing in target material, are finding wide use in television cameras. It seems a safe assumption that such tubes will assume an even greater role as time progresses.

The principal advantages of the new breed of camera tubes are related to the fact that the target is composed of semiconductor materials. While these tubes retain many of the principal advantages of the vidicon—small size, simplicity of adjustment, and stability—they also offer additional capabilities in areas where the vidicon is somewhat lacking. While the vidicon is quite sensitive, it cannot be called a fast tube because it tends to blur or smear fast moving objects in the image. This is particularly true at low

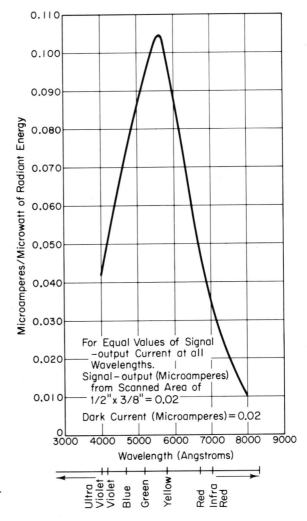

FIGURE 4-13. Typical spectral sensitivity characteristic.

levels of illumination where the vidicon is driven harder to obtain a higher sensitivity. The semiconductor target materials eliminate this effect to a large degree. They are also advantageous in the fact that they generate extremely small values of dark current.

The earliest commercially successful camera tube to employ semiconductor techniques was the "Plumbicon." ® It has become quite popular, especially in color television cameras where its low dark current, good sensitivity, and light transfer characteristics offer advantages. Its lag charac-

® Registered trademark of *North American Philips Company*.

teristics and lack of image retention are also attractive. The principles of operation are quite similar to that of the vidicon except that the resistive elements shown in the vidicon photoconductor (Figure 4-8) may now be thought of as semiconductor current sources that are controlled by light energy. The low dark current in those areas not illuminated is a result of the diode-like action of the photoconductor. When light is not present and target voltage is applied, the photoconductor elements are essentially reverse-biased diodes and the inverse current (dark current) that results is quite small. The tube has a disadvantage in that it is not a variable-sensitivity device and any adjustments to affect the sensitivity of the camera must be made to the optical system.

Other tubes employing semiconductor target elements have been developed and offer some exciting possibilities. Some of these are constructed using etching techniques (similar to that employed in the manufacture of integrated circuits) to obtain a very large number of discrete semiconductor elements over the face of the tube. Such tubes are presently limited in their resolution capabilities, but show great promise in many applications.

# 5

# SCANNING
# SYSTEMS

To transmit an image that is focused upon the photosensitive surface of a pickup tube it is first necessary to convert the optical information into electrical signals. Previous discussion has revealed how the beam in a pickup tube can discharge the target at any particular point at which it may be focused. It is necessary to utilize this small point to discharge the entire area of the scene which is optically focused on the face of the camera tube. Obviously it cannot do this while in a stationary position—it must be set into motion.

## THE SCANNING PROCESS

A descriptive analogy of the scanning process might be to consider the human eye as related to the printed page. With the eye stationary, relatively little descriptive information is transmitted to the brain, but if the eye is put into motion across the page it can, in effect, discharge the page of its information and send it to the brain. This familiar operation is sometimes referred to as scanning and it closely parallels the method employed in television.

All television systems use rectilinear scanning. That is, there are two separate scanning procedures occurring simultaneously, one moving the beam horizontally and the other moving the beam vertically. Both scans are linear; i.e., the movement of the beam, both horizontally and vertically, occurs at a constant rate of speed.

Figure 5-1 is a simplified illustration of the beam path over the photo-sensitive surface of a camera pickup tube. In a television system the picture tube in the monitor or receiver would duplicate exactly the beam excursions shown. As can be seen, the beam travels in both horizontal and vertical directions simultaneously.

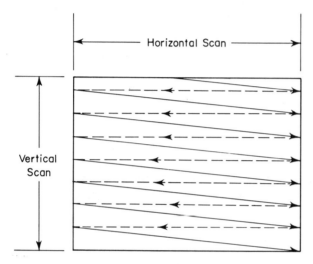

FIGURE 5-1. A simple scanning system. (Dotted lines indicate retrace.)

The dotted lines in the illustration serve to indicate *beam retrace*. Beam retrace occurs very rapidly with respect to normal scan and is normally *blanked* by disabling the beam in the cathode ray tube. Consequently, the camera tube develops no output signal during the retrace interval. It is also common practice to cut off the beam in the picture tube as well, thus blanking its retrace excursion and eliminating any visual evidence of the event.

In television the horizontal movement of the beam occurs at a much more rapid rate than the vertical movement. The horizontal therefore occurs at a higher *scanning frequency* than does the vertical. It is the relationship between these two frequencies (i.e., their numerical values) that determines the number of horizontal lines that will exist in each scanned pattern, or *raster*.

It is necessary to produce a relatively large number of horizontal lines per vertical scan to allow the beam to discharge a maximum amount of the target, thus yielding a greater amount of scene information. It is also desirable at the monitor picture tube because a greater number of scan lines reduces their individual visibility. It has been experimentally determined that the scanning lines should not be disturbingly visible to an average viewer at a distance of approximately six times the picture height when the display is used for entertainment or general observational purposes. At this distance the limitations of the human eye tend to blend the lines together if they are closely spaced.

If the television system is to be used to transmit images containing elements of fine detail, the number of scanning lines should be increased accordingly. Also, in instances where the viewer is quite close to the display tube (seated at a control console, for example), an increase in the number of lines will reduce their visibility.

### Flicker

A relatively low vertical scan rate is desirable because it allows the beam to inscribe a large number of horizontal traces without necessitating an inordinately high horizontal scanning frequency. However, the vertical rate cannot be too low or the visual presentation will exhibit an objectionable *flicker,* and be quite uncomfortable to watch.

Flicker varies considerably with picture tube illumination levels, viewing angle, and other factors aside from the frequency. Sensitivity to the effect also seems to vary between individuals. Therefore, to eliminate objectionable reactions, the vertical scanning frequency must be maintained sufficiently above that rate which is discernible to the average human observer.

The motion picture industry, with its film speed of 24 frames per second, has solved the flicker problem by causing each frame to be illuminated twice during the interval that it is being shown. This results in a flicker rate of 48 cycles per second and generally alleviates the problem. Such a rate, while acceptable for motion picture projection, can often be easily discerned on a television monitor. This is primarily due to the reduced display area and the relative brightness levels involved. Motion pictures are normally viewed in areas where the average light is subdued, while television displays may require very high brightness levels to overcome the surrounding ambient light conditions. In view of the foregoing, a vertical scanning frequency of 60 hertz was chosen as the standard in commercial broadcast television in the United States, and is generally used in closed circuit television systems as well.

It is important to note that the 60-hertz vertical scanning frequency is the same as 60-hertz power line voltage supplied by commercial power companies and used by most line-operated electronic equipment in the United States. This reduces the possible effects of equipment power supply ripple and 60-hertz magnetic fields in the final presentation on the monitor. This coincidence of frequencies also makes the line voltage a convenient source for synchronizing the vertical deflection frequency in simple scanning systems. The importance of these factors has diminished somewhat in recent years, but may be appreciated when it is recognized that countries that employ 50-hertz commercial power utilize a 50-hertz vertical deflection frequency as well. Such a low frequency would cause many American viewers to object to the rather borderline flicker effect, but European countries have long used such systems and their viewers have apparently accustomed themselves to the phenomenon.

### Interlace and the "Standard" Raster

Broadcast television in the United States uses a horizontal scanning frequency, or *line rate,* of 15,750 hertz. Simple division shows this to be 262½ horizontal lines of scan for each 60-hertz vertical deflection period. The raster thus produced by a single vertical scan is known as a *field,* and the vertical deflection frequency is referred to as the *field rate*.

Since the total number of lines per field (262½ ) does not constitute a whole number, it may be safely assumed that superimposing one field upon a succeeding field will not result in an exact mating of horizontal lines. In fact, since the number of lines dictate that each field begins ½ -line offset with respect to the preceding field, there occurs an interweaving or *interlacing* of horizontal lines on each alternate field. Consequently, if the horizontal lines of each of two succeeding fields fall directly between each other, the total number of horizontal lines will be effectively doubled to 525 (Figure 5-2).

Two interlaced fields comprise one *frame* whose repetition rate is 60/2, or 30 hertz. The frame constitutes a complete television picture element, but its low repetition rate does not generate flicker because of the two vertical scans that were required to complete it.

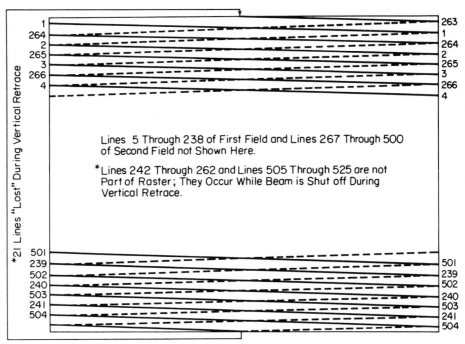

Details of Raster Produced by the 525-Line Scanning Pattern

FIGURE 5-2. Details of raster produced by a 525-line scanning pattern. Alternate lines are generated during alternate fields.

To reiterate:

$$\text{Line rate} = 15{,}750 \text{ hertz}$$

$$\text{Lines-per-field} = \frac{15{,}750}{60} = 262\frac{1}{2}$$

$$\text{Lines-per-frame} = \frac{15{,}750}{30} = 525$$

These figures describe the 2:1 interlaced "standard" scanning ratio of the United States.

The *aspect ratio* of the generated raster (its relative height vs. width) is 3:4. That is, it is 3 unit lengths high and 4 unit lengths wide (Figure 5-3). The additional width was undoubtedly chosen because most movement occurs in the horizontal plane, and a slight panoramic effect is also achieved.

In actual practice it is not possible to utilize all of the 525 lines of scan that are generated during each frame. About seven to eight per cent of such lines are rendered useless by the vertical retrace action and the blanking (beam cut-off) that accompanies it. This means that a total of 35 to 42 horizontal lines are lost during each frame (18 to 21 during each field). Therefore, the *active* or useful scanning lines will be reduced to between 483 to 490 (these are the limits of the tolerance set forth by standardized specifications).

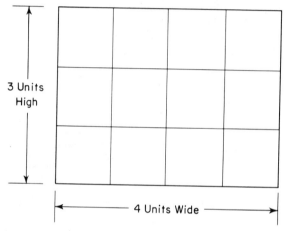

FIGURE 5-3. Standard aspect ratio.

### Vertical Resolution

In television, *vertical resolution* describes the ability of a system to resolve *horizontal* lines in a scene. The vertical resolution capability depends primarily upon the number of scanning lines that are used per frame. This is generally referred to as the *scan ratio* of a television system. It was previously stated that the "standard" system in the United States has a scan ratio of 525:1; however, closed circuit television systems have long used scan ratios of 1200:1 or higher. Since the horizontal lines must be crowded closer together to achieve these higher ratios, it follows that vertical resolution will increase accordingly.

It might at first seem logical to assume that a 525-line raster would yield 525 lines of vertical resolution, but this is certainly not so. There are several limitations that must be considered.

To begin with, the term *resolution* has a rather special meaning in television. In standard photography, for instance, four black lines separated by white lines of equal thickness would be defined as four lines of resolution. In television, the black lines *and* the white lines are *both* counted, giving a total of seven lines of resolution. Of course, this is simply a difference in definition, but it often causes confusion when comparing resolution capabilities of film vs. television.

To compute the vertical resolution capability of a specific scan ratio, it becomes necessary to introduce a modifying factor known as the *Kell factor*. To understand this term it is important to consider the finite size of the beamspot on the face of a camera tube in relation to a series of closely spaced horizontal black and white lines (Figure 5-4). In Figure

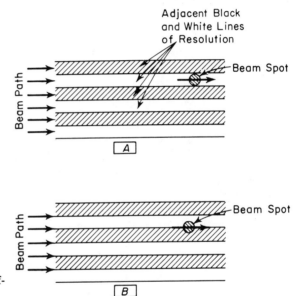

FIGURE 5-4. The beam spot size affects vertical resolution.

5-4(a) the spacing of the lines is such that there is one line of resolution for each horizontal scanning line. If the beam were perfectly aligned with each line it would seem that vertical resolution would indeed equal the number of active scanning lines. However, consider the results obtained in Figure 5-4(b) where the scanning lines have shifted slightly and the beam spot is always focused upon the junction of a black and white line. Obviously it cannot give a simultaneous output for both the black and white information. Instead, it integrates the effects of each and the resultant output falls somewhere between the two and is displayed on a monitor as a continuous grey, eliminating completely any evidence of the resolution elements.

Of course, the above examples are extreme, but they serve to illustrate the type of degradation that does occur. Experimentation by various agencies and individuals has yielded an "approximation of utilization," or Kell factor, of about 0.7.

Using the Kell factor to determine the vertical resolution that may be obtained from a 525-line system, it is merely necessary to multiply the 0.7 times the approximately 490 active scanning lines that remain after the vertical blanking action. This shows the vertical resolution to be approximately:

$$490 \times 0.7 = 343 \text{ lines of vertical resolution}$$

(Note: This figure is based on vertical blanking times used by most broadcast stations in the United States. Shorter blanking intervals, and the subsequent increase in active scanning lines, would increase the result accordingly.)

Most manufacturers of television equipment in this country specify vertical resolution of the standard scanning ratio at 350 lines. This is not an unrealistic figure since the Kell factor is only an approximation, and some less conservative estimates place it as high as 0.85. The variation in these estimates emphasizes the difficulty involved in accurately measuring the effect.

Utilizing the above information for a closed circuit television system employing a 1000:1 scan ratio yields:

$$0.07 \text{ (blanking time)} \times 1000 = 70 \text{ lines lost in blanking}$$
$$1000 - 70 = 930 \text{ active lines}$$
$$930 \times 0.7 \text{ (Kell factor)} = 651 \text{ lines of resolution}$$

Proper use of such simple mathematics will enable the prospective user to compute the scan ratio necessary for his application based upon specific vertical resolution requirements.

### Horizontal Resolution

In television, *horizontal resolution* is described as the ability of a system to resolve *vertical* lines of resolution. This ability is determined largely by the *bandwidth* of the television system (assuming the beam spot size in the pickup tube to be of extremely small dimension).

Consider the illustration of a series of vertical lines focused upon the target of a camera tube (Figure 5-5). As the beam traverses its horizontal path there is a series of output pulses generated by the beam action upon the target. The frequency of the pulses is directly proportional to the speed of the trace and the number of lines that it crosses. The video amplifiers in the camera system must have the capability of passing the highest frequencies thus generated or it is said to be *bandwidth limited,* and the horizontal resolution capability will depend upon the upper bandwidth limit.

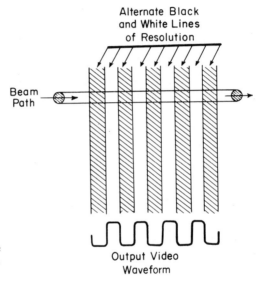

FIGURE 5-5. The speed at which the beam is scanned determines output frequency.

It should be noted here that all resolution measurements in television systems are made with respect to picture height. That is, *lines of resolution* would be defined as that number of black and white lines of equal width that could be contained in the vertical dimension of the picture. This is also true when speaking of horizontal resolution, even though the horizontal dimension is greater because of the 4:3 aspect ratio. Therefore, when it is said that a system has a 500-line horizontal resolution capability, it means that it can resolve resolution elements of such width that 500 of them, placed side by side, would exactly fill the vertical dimension of the television picture.

Calculation of horizontal resolution is slightly more involved than the procedure for determining vertical resolution capability, but simple mathematics still apply.

Assume that a need exists for a 600-line horizontal resolution capability in a standard 525-line system. Remembering that a picture element, or line of resolution, as used in television, may be either black or white, a series of four black lines with four alternate white lines of equal width represents eight lines of resolution. However, the electrical signal generated by a vidicon viewing such lines will represent an adjoining black and white pair as a single cycle (Figure 5-5), producing 300 cycles for a 600-line resolution capability.

Because the horizontal frequency of the system in question is 15,750 hertz, one horizontal line is generated in 1/15,750th of a second, or 63.5 microseconds. However, horizontal blanking time occupies 16 to 18% of this period, leaving approximately 53 microseconds of useful time in the single horizontal line. Also, since the 600 lines of resolution are specified in relation to picture height, a modifying factor of ¾ must be introduced to compensate for the 4:3 aspect ratio. This leaves about 40 microseconds of usable time in the horizontal line.

Using the above information to calculate the bandwidth required to present the 600 lines of resolution yields:

$$600 \text{ lines} = 300 \text{ cycles of video signal}$$

$$300 \text{ cycles in 40 microseconds} = 300 \times \frac{1}{40 \times 10^{-6}} = 7.5 \text{ megahertz}$$

It may therefore be concluded that 7.5 megahertz is necessary to present the horizontal resolution of 600 lines in the standard scanning format.

Conversely, resolution may be computed when bandwidth is known. Using the 4.0-megahertz bandwidth of broadcast television in the United States:

$$4.0 \text{ megahertz} \times 40 \text{ microseconds} = 160 \text{ cycles}$$
$$160 \text{ cycles} \times 2 = 320 \text{ lines of resolution}$$

And, using a 10-megahertz bandwidth:

$$10 \text{ megahertz} \times 40 \text{ microseconds} = 400 \text{ cycles}$$
$$400 \times 2 = 800 \text{ lines of resolution}$$

Examination of the above equations will reveal that the standard 525-line scanning format will produce about 80 lines of resolution for each megahertz of bandwidth.

### Scan Rate vs. Bandwidth

When a television system utilizes a scanning ratio that is greater than the standard 525:1, the beam must move faster to achieve the greater ratio. For example, if a scan ratio of 1000:1 were used, the beam would have to move almost twice as fast as before to create 1000 scan lines in the time that it previously took for 525. (Remember that the vertical frequency remains constant at 60 hertz in all cases.)

With the beam moving faster, the frequencies that it generates when crossing lines of resolution will be increased accordingly. Therefore, to achieve an 800-line horizontal resolution capability with a scan ratio that is greater than the previous 525:1 the bandwidth will have to be increased.

In the case of a 1000-line system:

Horizontal rate $= 30,000$ hertz
$1/30,000 = 33.3$ microseconds horizontal interval
Active time (less blanking) $= 28$ microseconds
$28$ microseconds $\times$ ¾ $= 21$ microseconds
$400$ cycles $\times$ $\frac{1}{21}$ $\times$ $10^{-6}$ $=$ slightly over 19 megahertz

Thus, presentation of 800 lines of vertical resolution with a system utilizing a 1000:1 scanning ratio would require a bandwidth of approximately 20 megahertz.

The foregoing examples should serve to demonstrate the rigid relationship that exists between resolution capability, scanning ratio, and bandwidth. Although it might initially appear that extremely wide bandwidths are advantageous regardless of the scanning ratio, it should be pointed out that excessive bandwidth serves only to amplify noise. The scan ratio and bandwidth should therefore be chosen to fill a particular requirement.

## METHODS OF DEFLECTION

There are two methods by which the beam in a camera tube or picture tube may be deflected. These methods depend upon the properties displayed by charged particles (in this case, electrons) that are moving within electrostatic or electromagnetic fields.

**Electrostatic Deflection**

Electrostatic deflection utilizes characteristics exhibited by electrons when placed between two conductive plates that have electrical charges of opposite polarity impressed across them. Following the universal law that like charges repel each other and unlike charges attract, the electrons, being negatively charged particles, will be attracted to a positively charged plate and repelled from a negatively charged plate.

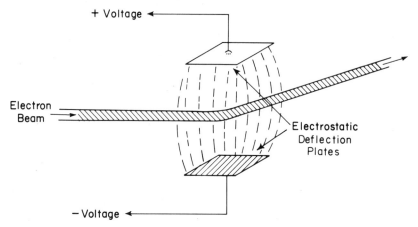

FIGURE 5-6. Electrically charged plates deflect an electron beam.

If an electron beam is passed between two plates of opposite charge (Figure 5-6) there is a decided deviation of the beam from its original path. The amount of deviation is dependent upon electron mass and velocity, and the strength of the electrostatic field (which is a direct function of the amount of voltage applied to the plates). The electrostatic method of deflection requires voltages of a relatively high order, but draws little current.

Electrostatic deflection is used in television systems less often than electromagnetic deflection because the electron beam tends to become defocused as the angle of deflection increases, and correction of this tendency becomes complex and costly. However, there are vidicon tubes that employ electrostatic deflection to good advantage.

### Electromagnetic Deflection

Most television cameras and display monitors employ magnetic energy as the principal means of causing beam deflection. The magnetic field generally allows maintenance of good beam characteristics and can be electronically manipulated to achieve deflection.

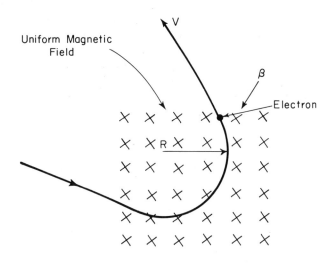

FIGURE 5-7. Effect of a uniform magnetic field upon a moving electron.

$\beta = $ Flux Density
$R = $ Radius of Curvature
$V = $ Velocity

Figure 5-7 illustrates the behavior of a moving electron in a uniform magnetic field. The small $x$'s in the diagram denote magnetic lines of force with a directional characteristic away from the viewer. (The $x$'s can be thought of as fins on arrows directed into the page.) The path of the electron describes a partial circle, with the particle moving always perpendicular to the lines of magnetic flux. The radius of the orbit is determined by the mass of the particle, its velocity, and the flux density (magnetic field strength).

Figure 5-8 describes the action of a magnetic field upon a stream of electrons that constitutes the beam of a cathode ray tube. It may be noted that the electrons depart from their original course with a circular deviation while they are within the magnetic field. The radius *R,* determined by the factors previously mentioned, establishes the amount that the beam will deviate from its initial path. Once the electrons are free of the magnetic field, the beam path again becomes straight.

It is relatively simple to visualize the effect of gradually reducing the strength of the magnetic field and then reversing it. As the field weakens, the amount of deflection becomes less and, upon field reversal, the beam is forced to the other side of its original path. A field that alternates repetitively in this manner will cause the beam to scan back and forth.

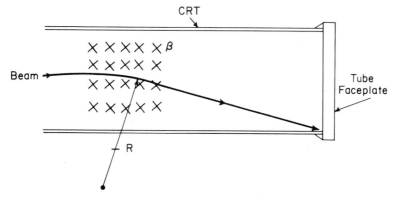

FIGURE 5-8. Electromagnetic deflection in a cathode ray tube.

In order to accomplish the above, it becomes necessary to use *electromagnetic* energy. Yokes, composed of precisely wound coils of wire, are placed around the cathode ray tubes and the scanning process is achieved by passing controlled amounts of current through them to generate the magnetic field.

The ideal way for a beam to scan is for it to move at a uniform rate of speed in a straight line across the target area. Reaching the edge of the area of interest, it should move quickly back across the target (retrace) and begin a new trace at a point just slightly lower than that of the previous trace. Repeating this procedure until it reaches the bottom of the raster area, the beam should then retrace to the top and begin another vertical sweep.

The sawtooth current waveform necessary to perform a linear sweep is shown in Figure 5-9. The gradual decrease in the current waveform from the negative direction allows the beam to return toward the center of the target area. When the waveform reaches that point on the ramp where the current is zero, the beam is in the center of its trace. As the current waveform goes positive, deflection to the other side of center takes place. The sharp, negative-going edge of the sawtooth waveform causes the beam to quickly retrace and begin another scan. It should be noted that if the ramp of the sawtooth waveform is not a perfectly straight line, but is bowed as shown in dotted lines, the speed with which the beam sweeps across the target will not be constant.

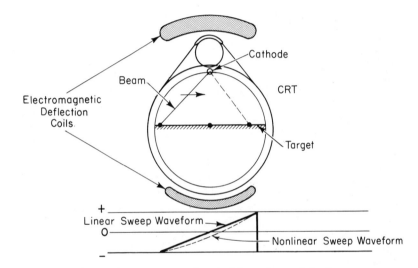

FIGURE 5-9. Sawtooth current waveform achieves beam deflection.

## LINEARITY AND SWEEP DISTORTION

The rate at which the beam moves across the face of an image pickup or display tube must be constant or the sweep is said to be nonlinear. A nonlinear scan will cause certain areas of the resultant display to be compressed or expanded relative to the original optical image that was focused on the pickup tube. Since there are two sawtooth current waveforms necessary to form a complete raster with both horizontal and vertical dimensions, it follows that nonlinearity is possible in either of these dimensions.

There are many things that can make a display nonlinear; so many in fact that it is virtually impossible to compensate for them to the extent

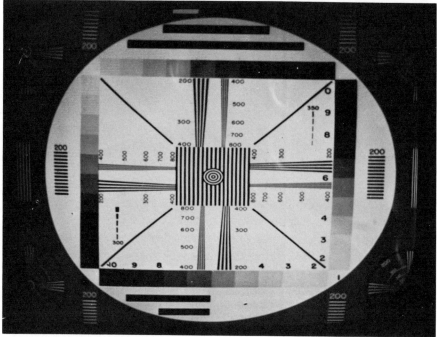

FIGURE 5-10.  Examples of nonlinear deflection systems.

that perfect linearity is achieved. In actual practice, an electromagnetically deflected display that deviates in linearity less than plus-or-minus one per cent of picture height is considered to be more than adequate for most purposes.

It is usually true that precise horizontal linearity is more difficult to accomplish than is vertical linearity. This is caused by difficulties with the higher horizontal frequency as associated with the inductive deflection coils and associated components.

The photographs in Figure 5-10 illustrate some rather exaggerated examples of nonlinearity. While these are admittedly extreme cases, it should be pointed out that the human eye is a critical judge of linearity when provided adequate references. A system with moderate linearity deviations may be used to display a scene of general content without being objectionable to the average viewer. However, if the same system is used to view a circle, for example, the problem becomes glaringly obvious. A series of parallel lines or a checkerboard pattern also offer good visual checks on system linearity.

Because a television system contains deflection assemblies in both the camera and the monitor, nonlinearity can be introduced by either device. Initially it might appear difficult to determine where system non-linearity might be originating, but independent consideration of the camera and monitor linearity functions reveals an interesting difference between the characteristics of the two.

### Camera Nonlinearity

Nonlinearity in a television camera results in an output video signal that represents image elements improperly with respect to time. To illustrate, consider a camera focused upon a series of parallel vertical lines spaced an equal distance apart (Figure 5-11). In this case, let's say that the beam moves more slowly as it scans the left side of the scene and speeds up as it crosses the right side. Because scanning with a beam gives a signal output that is proportional to the number of lines and the speed of the scan across them, the nonlinear scan will cause a deviation in the frequency of the video output signal, as shown in the diagram. Displaying this signal on a linear monitor will result in a pattern whose lines are unevenly spaced.

To provide an accurate check of camera linearity it is necessary to add an externally generated signal to the video signal generated by the camera. If the frequency of the external signal is adjusted properly (and assuming the signal to be narrow, positive-going pulses), a series of narrow white lines will be displayed on the monitor. Because of the timing accuracy of the generator these lines will occur at accurate intervals with respect to time.

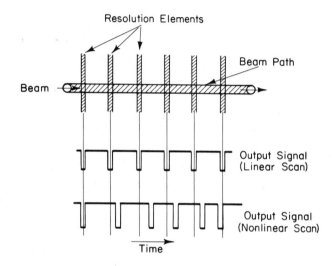

FIGURE 5-11. Nonlinear scan of evenly spaced resolution elements results in frequency distortion of output waveform.

If the camera is used to view a series of equidistant vertical stripes, the output signal *should* be a series of accurately spaced pulses. Comparison of these pulses with an equal number of known accurate pulses from the external generator would establish the existence of any nonlinearity in the camera. Adding the two signals and displaying them on a monitor provides an excellent means of comparison. This establishes and unalterable relationship between the camera and generator signals. Thus, any displacement that exists between them will show up on a monitor regardless of the linearity characteristics of the monitor itself. This procedure is covered more fully in Chapter Nine.

### Monitor Nonlinearity

Monitor nonlinearity, like camera nonlinearity, is caused by a variation in the speed of the scanning beam as it traverses the face of the CRT. Monitor nonlinearity may be detected by capping the lens of the camera so that no video information generated by the camera scanning system is present and then feeding the signal from the external pulse generator into the monitor. This will cause lines to appear on the monitor that are of a known constant frequency, and a perfectly linear monitor should show them as being equally spaced as viewed on the screen. The distance between the lines can be measured to check the spacing, or an overlay or projection with an equally spaced pattern may be used for direct comparison.

### Geometric Distortion

Linearizing circuits are incorporated in both the camera and monitor to allow control over the shape of the deflection waveforms. However, it is possible to have near perfect linearity of these signals and still have some extreme problems in achieving a uniform display. The major factor in such problems may be found to be within the deflection yoke itself, and the irregularities thus originated are usually referred to as *geometric distortion.*

Geometric distortion is often caused by an improper relationship of the windings of the yoke. By way of example, Figure 5-12 illustrates a type of geometric distortion known as *yoke skew*. It is caused by the improper positioning of the horizontal windings in the yoke with respect to the vertical windings. They should be positioned around the tube at a 90-degree angle relative to each other. If it is possible to conveniently adjust this relationship (in most cases it is not), the problem can be cor-

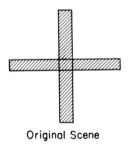

Original Scene

Monitor   Presentation

FIGURE 5-12. Yoke skew introduces angular distortion of horizontal-to-vertical relationships.

rected. It is unlikely that this display would be acceptable for most applications even if the linearity waveforms were perfect.

There are other types of geometric distortion which commonly occur. A compressing or stretching of one corner of the raster is not unusual, and can usually be traced to a yoke imperfection which causes a non-symmetrical magnetic field. Similarly, external magnetic fields that are strong enough to penetrate the camera or monitor shielding may have an effect upon the electron beam displacement and yield a form of geometric distortion.

## SCANNING CIRCUITRY (CAMERA)

The vertical and horizontal scanning circuits of a camera system must provide sawtooth current waveforms to the deflection coils of the yoke. Such waveforms must be of the proper amplitude for driving the particular yoke that is used and of course be linear in nature. Since the vertical deflection angle is less than the horizontal, and its frequency is considerably lower, less power is consumed by the vertical circuitry. Vertical circuitry is also generally simpler in design than the horizontal.

### Vertical Deflection Circuits

There are several methods that may be used to generate vertical deflection waveforms. Figure 5-13 illustrates the use of a unijunction transistor oscillator.

Unijunction relaxation oscillators are often used as sources of vertical

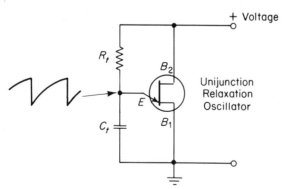

FIGURE 5-13. Sawtooth generation by a relaxation oscillator.

deflection because they are inherently simple circuits that yield a respectable sawtooth waveform. The unijunction has one emitter and two bases. Base 2 is connected to a positive source and, in this case, base 1 is routed to ground. With a specific supply voltage, there is a value of positive voltage which, if applied to the emitter, will cause the emitter-base 1 junction to become forward biased. Prior to such time the resistance between the two points is quite high. When forward biased, emitter-to-base 1 resistance is reduced to a minimal value and relatively large currents can flow through the junction. The value of voltage at the emitter that will cause such conduction is a function of the voltage that is applied at the *base 2* terminal. When power is initially applied, the emitter is reversed biased and not conducting. Capacitor $C_t$ will begin to charge toward a positive potential through $R_t$ and will continue to do so until reaching the point at which the emitter-base 1 junction becomes forward biased. When this occurs the capacitor will discharge through the junction, abruptly lowering the voltage at the emitter. When the emitter voltage becomes so low that it cannot sustain conduction, the emitter-base 1 junction again becomes reverse biased and the cycle begins to repeat itself. The resulting waveform at the emitter is a nonlinear sawtooth whose frequency is determined by $R_t$, $C_t$, and the characteristics of the unijunction device.

Since the point at which the unijunction conducts is determined by the voltage at the emitter, it becomes a relatively simple matter to lock the device to an external frequency by adding positive-going pulses at the emitter. These pulses add to the potential that is building up in the capacitor and cause the emitter-base 1 junction to become forward biased prior to the time such an event would normally have occurred. Vertical oscillators are often triggered in this manner by external synchronizing generators.

Figure 5-14 is a more complete diagram of a vertical oscillator, followed by an emitter follower whose output is employed as feedback for linearizing the sawtooth waveform. The final stage is another emitter follower whose output amplitude is controlled by the *vertical size* potentiometer. The output from this stage is routed directly to the yoke to achieve the desired vertical beam deflection within the camera tube (in this case, vidicon).

*Vertical centering* is achieved by passing a dc current through the yoke to establish the point at which the raster will be centered.

Another method of sawtooth generation for vertical deflection is illustrated schematically in Figure 5-15. Here a negative trigger pulse

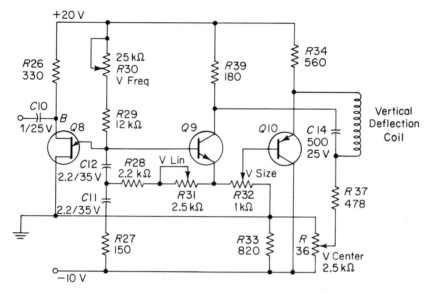

COURTESY COHU ELECTRONICS, INC.

FIGURE 5-14. Vertical deflection circuit.

occurring at the vertical deflection rate drives $Q1$ into saturation and provides a positive pulse for the sawtooth input circuitry. The positive pulse thus generated causes the voltage across $L1$ to go positive in a sinewave excursion through the conducting diode $CR1$, at a frequency dependent upon $L1$ and $C1$. The frequency of the sinewave is quite high, making the rise time to the positive potential relatively fast. When the positive excursion begins to reverse itself, $CR1$ becomes reverse biased and ceases conducting, leaving $C1$ positively charged at its connection to the base of $Q2$. The capacitor then begins a slow $RC$ discharge through $R3$ and height potentiometer $R4$, creating a negative sawtooth waveform. Transistor $Q2$ is essentially an emitter-follower amplifier, but it is "bootstrapped" from the emitter back to its base through $C2$ to reduce transistor loading. The sawtooth output is also taken from the emitter which is a point of low impedance.

Transistor $Q3$ operates as a comparison amplifier and compares the output of the sawtooth generator with the ac feedback from $Q4$, the output stage. The feedback waveform is obtained by driving a sample of the yoke current through a sampling resistor, $R10$. $Q3$ then amplifies the error signal between base and emitter, caused by the differences in the two waveforms, and corrects any nonlinearity in the scan.

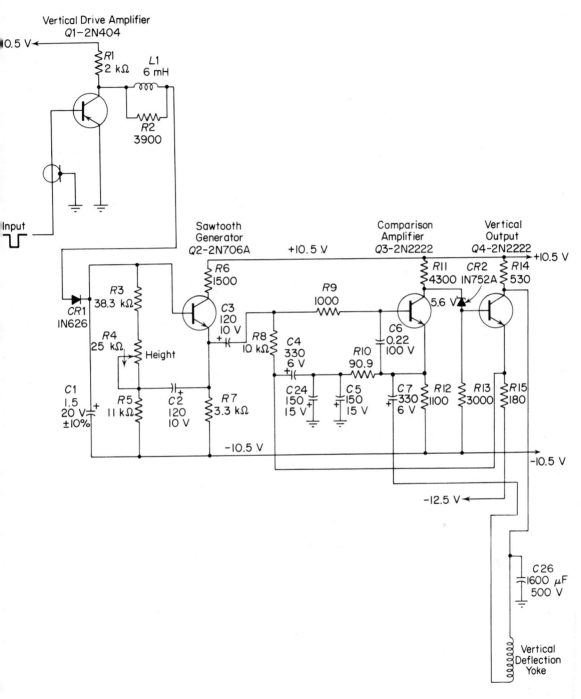

FIGURE 5-15. Another vertical deflection circuit.

### Horizontal Deflection Circuits

Deflection circuits associated with the horizontal scanning function are considerably different from those used for vertical deflection.

Figure 5-16 is a simplified schematic of one method used to deflect the beam horizontally in a vidicon camera. The voltage across the yoke does not appear as a sawtooth in this case. The current waveform, however, does constitute a sawtooth. This fact may be realized by placing a small value resistor in series with the yoke and observing the voltage waveform that is developed across it. It is the inductive load that causes the difference between voltage and current waveforms. The high-amplitude, negative-going voltage pulse shown is necessary to offset the increase in reactance that results from the increase in frequency rate noted with respect to the vertical deflection.

In operation, the transistor $Q$ serves as a switch in the coil and power supply network. If the transistor is switched on, the current through the yoke begins to increase slowly in a sawtooth manner. If, at the beginning of each horizontal interval, a drive pulse turns the transistor off for a brief period, all current flow into the yoke ceases. The abrupt loss of current drive to the yoke causes the magnetic field associated with it to collapse, reversing the current and causing an inductive kick that is impressed on the diode, transistor, and capacitor. The diode and transistor are, of course, nonconductive at this instant, but the yoke and capacitor form a tank circuit and will tend to oscillate or "ring." The first half-cycle results in a high-amplitude negative pulse. However, when the oscillation tries to go positive, the diode conducts and shorts the capacitor.

The current through the yoke reverses direction during the short, half-cycle interval, causing the beam in the tube to reverse its direction of travel, or *retrace*. Because of the sudden change in current flow, the retrace occurs very rapidly with respect to the normal scan. The reverse current that has thus been established through the yoke slowly decreases toward zero, forming the first half of the sawtooth waveform and the first half of the beam's horizontal trace. When the circuit reaches the point where the yoke can no longer supply current to keep the diode in conduction, the transistor again turns on, causing current to begin flowing in the opposite direction, thus continuing the sawtooth current waveform through the zero point and causing the beam to progress beyond the center of the screen and complete the last half-cycle of its trace.

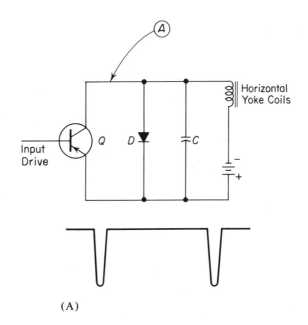

(A)

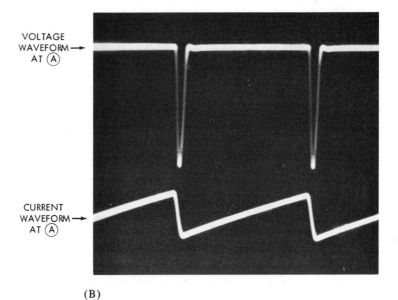

(B)

FIGURE 5-16. Horizontal deflection using "flyback" principles. Voltage and current waveforms differ considerably.

Another drive pulse applied at the input to the transistor causes the cycle to repeat itself, giving a sawtooth waveform output occurring at the horizontal rate. It should be remembered that the waveform here is a *current* waveform, and cannot be directly observed with an oscilloscope. The oscilloscope would simply see the high amplitude, negative-going retrace pulse if a probe were attached to the yoke.

Figure 5-17 schematically illustrates a circuit that utilizes the above principles to achieve horizontal deflection. In this instance, $Q1$ is simply a pulse shaping stage used to supply a drive pulse of the necessary amplitude to the horizontal output transistor $Q3$, through emitter follower $Q2$ which functions to provide impedance matching and current amplification.

Transistor $Q3$ is normally in a conducting state and fully saturated due to the biasing action of resistor $R5$ at its base. Application of a positive-going pulse to the base will cause it to turn off. The abrupt loss of current to the yoke causes the retrace pulse to be generated, initiating beam retrace. The positive-going transition of the resultant ring caused by the tank circuit is clipped by the action of diode $CR1$, which becomes positively biased when the amplitude of the output pulse returns to a $+20$ volts.

The resonant frequency of the tank circuit is an important factor to be considered. The frequency at which it resonates will determine the pulse width of the horizontal retrace pulse. This pulse must be kept reasonably narrow so that beam retrace may be completed in the relatively short interval allowed for it. $L2$ is a series inductor included to linearize the sawtooth current through the yoke. In this example, a saturable reactor is used which allows some control of scan linearity.

Control of horizontal size is achieved by providing control over the amount of voltage felt at the collector of $Q3$. It is varied by potentiometer $R17$, the *horizontal-size* control. $L1$ and $C4$ act as a filter network, keeping the horizontal output pulse from affecting the horizontal size control.

Horizontal centering may be controlled by varying potentiometer $R20$. This determines the amount of dc current that will be routed to the horizontal yoke through isolation choke $L3$, thereby positioning the sweep pattern horizontally on the vidicon face.

Figure 5-18 shows horizontal deflection circuitry which is very similar to that of the vertical deflection circuit of Figure 5-15. The primary differences are in component values to accommodate the higher horizontal frequency. Choke $L4$ is included to provide low impedance to dc elements and a high impedance to the ac (horizontal frequency) elements.

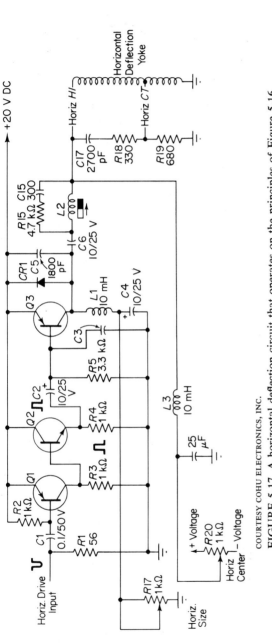

FIGURE 5-17. A horizontal deflection circuit that operates on the principles of Figure 5-16.

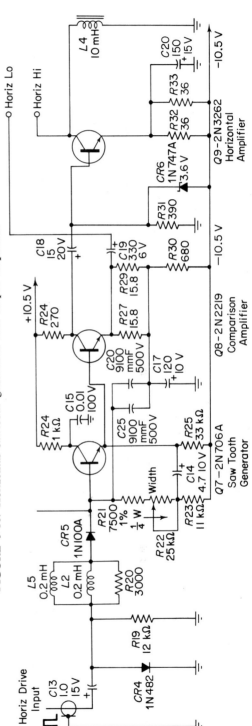

FIGURE 5-18. Horizontal sawtooth generation and output amplifiers.

### Horizontal Delay Circuits

Many television systems employ cameras (or camera heads) that are located a considerable distance away from the video processing circuitry. Because sync and blanking waveforms are added to the video waveform in the processing circuits and not in the camera head itself, timing problems can arise. If a camera is located a couple of thousand feet from the processor, as is often the case in many closed circuit systems, the time involved in running horizontal drive pulses up the cable and the resultant video signal back down the cable can be appreciable. The delay can be so extensive that the blanking and sync could be added as shown in the exaggerated example of Figure 5-19. When the sync, blanking, horizontal drive, and vertical drive arrive at the processor they all possess the proper time relationship with each other. However, the delay experienced by the horizontal drive going to the camera head will cause the horizontal deflection in the camera head to be late in starting. This means that the video signal

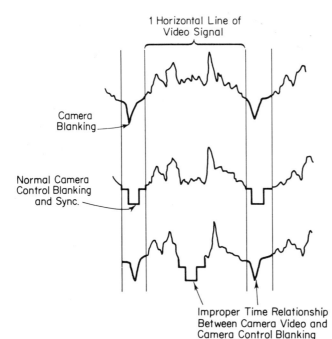

FIGURE 5-19. Improper time relationship between signals renders the video signal useless.

developed by that particular scan will not bear the proper time relationship to the blanking and sync that are to be added at the processor. The video signal will experience another equal delay in the cable going to the processor.

From the above it can be seen that a need exists for some means of advancing the horizontal drive pulse in time by an amount equal to the delays in the circuit. Of course, it is not possible to advance a pulse ahead of the one from which it was derived, so the horizontal circuits employ a delay network that essentially delays the output pulse one full horizontal line. Therefore, the pulse that will generate a horizontal line of video is derived from the horizontal drive pulse of the line before. The delayed pulse can therefore be adjusted so that it *appears* to advance with respect to the blanking and sync signals.

Figure 5-20 illustrates, in simplified form, circuitry that will accomplish the horizontal advance or, more properly, delay function. Transistor $Q1$ accepts the pulse from the synchronizing source and acts as an isolating stage because of its emitter-follower configuration. This stage triggers a cable delay multivibrator, composed of $Q2$ and $Q3$, into its unstable state, where it remains for a period of time determined by the values of $R7$ and $C4$. The duration it remains in the unstable state may be controlled by

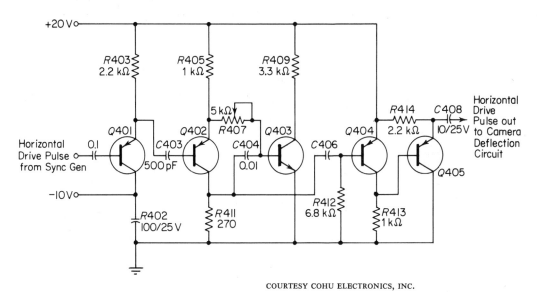

COURTESY COHU ELECTRONICS, INC.

FIGURE 5-20. A pulse delay circuit.

adjustment of potentiometer $R7$, which determines the width of the output pulse generated.

The trailing edge of the multivibrator output pulse is fed to the pulse-shaper stage $Q4$. With no signal at $C6$, $Q4$ conduction is cut off and remains off until $C6$ is charged through $R12$ and the transistor is again biased into saturation. The length of the nonconduction period is determined by the $R/C$ time constant set by $R12$ and $C6$, which establishes the output pulse width.

The output of $Q4$ is fed to emitter follower $Q5$ for current amplification and impedance matching to drive the pulse through the cable. When $R7$ is adjusted so that the multivibrator output is equal in time to one horiontal line, the regenerated drive pulse is coincident with the next input pulse and no delay or advance is effected.

When $R7$ is adjusted for less than one horizontal line period, the regenerated pulse appears to occur before the input pulse. Advancing the regenerated pulse with reference to the input pulse causes the camera deflec-

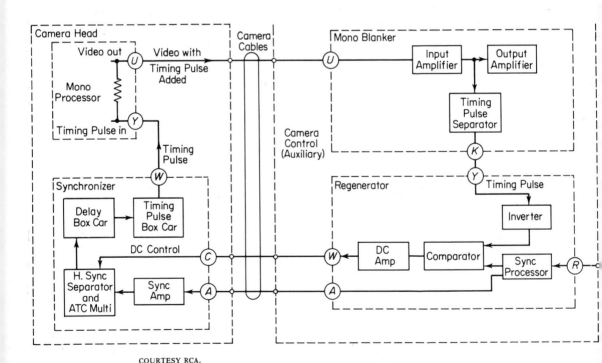

COURTESY RCA.

FIGURE 5-21. An automatic delay system.

tion circuit to be triggered ahead of the blanking and sync pulses occurring in coincidence with the input horizontal drive pulse. Again, the time difference is used to compensate for the delay of the outgoing pulse and the returning camera video, to maintain the proper time relationship to the sync and blanking pulses as inserted at the processor.

Figure 5-21 is a block diagram of a horizontal advance system that automatically adjusts itself for any length of cable between the camera and its auxiliary control unit. It does this by using the horizontal pulse that arrives at the camera to generate a test pulse that is later added to the video. The test pulse is of very short duration, (approximately 1 microsecond) and it is added to the camera video so that its trailing edge coincides with the leading edge of the camera blanking excursion. The video is then routed back to the auxiliary control, experiencing a delay proportional to the length of the cable. Thus, the test pulse arrives at the auxiliary control with a time delay equal to that associated with *twice* the length of the cable (the delay of the sync pulse to the camera, as well as that of the returning video). The trailing edge of the delayed timing pulse is compared with the leading edge of the horizontal sync in the auxiliary control and a dc control voltage is generated. The control voltage is sent back to the camera where it adjusts the period of the Automatic Timing Control multivibrator. The output of the ATC multivibrator reacts to the dc voltage by adjusting itself in a manner that advances the drive to the horizontal deflection until the timing pulse and sync pulse in the auxiliary control achieve their proper timing relationship. This causes the video out of the auxiliary control to be properly timed with the incoming sync.

Figure 5-22 is a schematic illustration of the Automatic Timing Control multivibrator from the previous example. The circuit is triggered by a drive pulse obtained by differentiating the sync pulse with capacitor $C15$ and resistor $R30$. The leading edge of the positive pulse obtained passes through diode $CR10$ to trigger $Q11$. $Q11$ and $Q12$ form a monostable multivibrator whose output-pulse width is determined by $C13$, $R27$, and the dc control voltage applied at the junction of $R27$ and $R28$. As the control voltage goes more positive, the output pulse width becomes shorter, and the more negative the control voltage becomes, the wider is the output pulse. It is this output pulse width (specifically, its trailing edge) which ultimately determines the timing of the horizontal deflection by providing the advanced horizontal drive pulse, and regulates the positioning of the timing pulse on the camera video waveform.

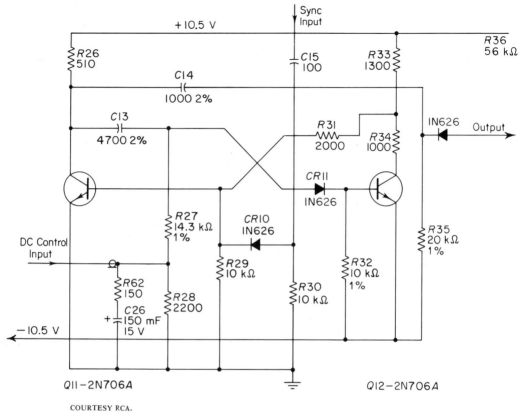

FIGURE 5-22. DC input controls output pulse width.

### Vidicon Blanking and Protection

It is necessary to blank the camera tube during horizontal and vertical beam retrace to prevent the beam from prematurely discharging target elements that are to be scanned during the next field interval. If blanking were not accomplished, the discharged areas would not be able to sufficiently re-establish enough charge to achieve their proper output, and dark retrace lines would degrade the presentation.

Many blanking circuits perform the dual role of protecting the camera tube from the effects of scan failure. If either the horizontal or vertical scan waveforms should be lost, the beam would continually retrace a single horizontal or vertical line, damaging the target area of the tube and leaving a permanent scan burn. By providing a means of sensing the presence of horizontal and vertical drive pulses, it becomes possible to disable the beam whenever either waveform is lost. Blanking and protection can be accomplished by driving the camera tube control grid negative, or the cathode positive, until the beam is cut off.

Figure 5-23 illustrates circuitry that develops cathode blanking signals for a vidicon tube. It is the function of $Q5$ and $Q6$ to act as switches, coupling a positive voltage to the vidicon cathode whenever the negative-going transitions of either horizontal or vertical deflection waveforms are present. Both transistors are normally in the off condition, turning on only when the horizontal and vertical retrace intervals dictate vidicon blanking. The presence of inputs, vertical at the base of $Q5$ and horizontal at the base of $Q6$, also causes a positive bias to be developed at both bases, which keeps the transistors off during the interval between pulse inputs. However, in the absence of either deflection signal, the corresponding transistor will become biased into the on condition, coupling a continuous positive voltage to the vidicon cathode, cutting off the beam current and thus protecting the vidicon target from damage. Additional protection is afforded by $Q4$ which senses the positive cathode voltage during protection and turns on, shorting the vidicon target voltage to ground.

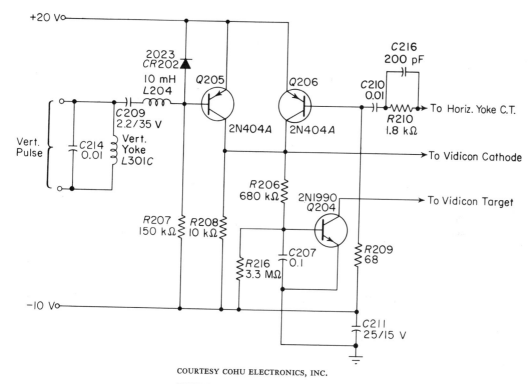

FIGURE 5-23. Vidicon blanking and protection circuit.

# 6

# VIDEO AMPLIFICATION
# AND PROCESSING

The video signal that is derived from the television camera tube must be amplified and processed before it is suitable for use in a monitor or receiver. In a vidicon tube, for instance, the output signal is measured in microvolts, while the standard video signal in closed circuit television is 1 volt. Amplifiers, whose gain is constant over a wide range of video frequencies, must be used to achieve the necessary increase in signal. Also, because of other characteristics of the camera tube, the signal must be processed to compensate for camera tube gamma characteristics, frequency response, and output distributed capacity. Likewise, the signal must be clamped to a stable dc reference voltage, and blanking and sync signals must be added. Other processes may or may not be performed depending upon the application for which the camera is intended and its relative sophistication.

Generally, the initial stages of amplification in a television camera provide almost all of the system gain, bringing the signal level up to between 0.5 and 1.0 volt before the processing stages are encountered. This is necessary to provide a good signal-to-noise ratio. If the processing were performed on low-level video signals, the noise generated in the various stages would constitute an appreciable portion of the signal. Therefore, amplifying this signal in later stages would increase the amplitude of the previously generated noise as well as the video.

Early amplification of the video signal is mandatory when a two-unit camera system is employed. The long lengths of cable that may separate the camera head from the processing circuits necessitates rather large signal amplitudes at the output of the camera head. In these instances a *preamplifier* is placed in the camera head, and the *processing amplifier* is located in the camera control unit. It is the function of the preamplifier to provide impedance matching for the vidicon or other camera tube, effect the major portion of gain for the system, and provide impedance matching into the cable that goes to the camera control unit. The processing amplifier then performs most of the additional functions necessary to condition the signal for display.

## VIDEO PREAMPLIFIER

Figure 6-1 illustrates an example of a video preamplifier circuit that was designed for use in a commercial broadcast camera. The input transistor, $Q5$, is located as near to the vidicon output as practical. Close proximity to the signal source minimizes the input circuit capacitance and also reduces the noise and RF pickup problems that might be caused by relatively long connecting wires between the vidicon target and the transistor input.

$Q5$ is an emitter follower, providing a high impedance input to match the vidicon output and acting as a low impedance source to drive the succeeding transistor stages. $Q2$ is another emitter follower, which lowers the impedance still further. The output of $Q2$ drives a *feedback pair* consisting of $Q3$ and $Q4$. The preamplifier gain is achieved in these two transistors and is determined by the amount of signal that is applied as feedback to the base of $Q3$ through $R8$. The $Q4$ emitter has a low impedance output which is coupled through the adjustable peaking network, $R13$ and $C9$, to match the low impedance load of the succeeding processing stages. $Q1$ also receives drive from $Q2$, which it applies to the collector of the input transistor, $Q5$, as feedback, routing it on the shield of the coaxial cable which carried the input signal, thereby cancelling the capacitive loading effect on the transistor elements.

The peaking elements, $R13$ and $C9$, compensate for the effects of distributed capacity at the vidicon target and the conductor from it to the preamplifier, etc. The small amount of distributed capacity encountered affects the signal appreciably because of the high output impedance of the vidicon.

Figure 6-2 shows a video preamplifier which has a 10-megahertz bandwidth. The input from the vidicon is peaked at the higher frequencies

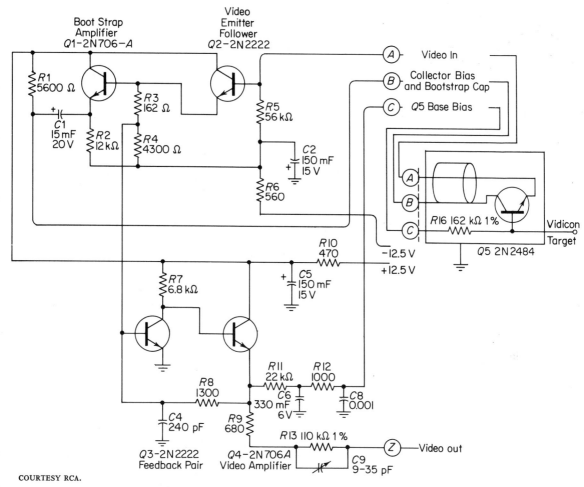

COURTESY RCA.

FIGURE 6-1. A video preamplifier circuit.

by the action of the input network consisting of $L101$ and $R102$. The signal is developed across $R104$ and coupled into $Q101$ by $C102$. $Q101$ is an emitter follower, which drives a feedback pair, $Q102$ and $Q103$. The feedback pair provides about 40 decibels of gain, and feeds the emitter-follower output stage, $Q104$.

Figure 6-3 incorporates parallel-connected FET input amplifiers and an integrated circuit (shown blocked in dotted lines). The high gain and low noise characteristics of the FET's, coupled with their high input impedance, makes them an especially effective input stage. The output from

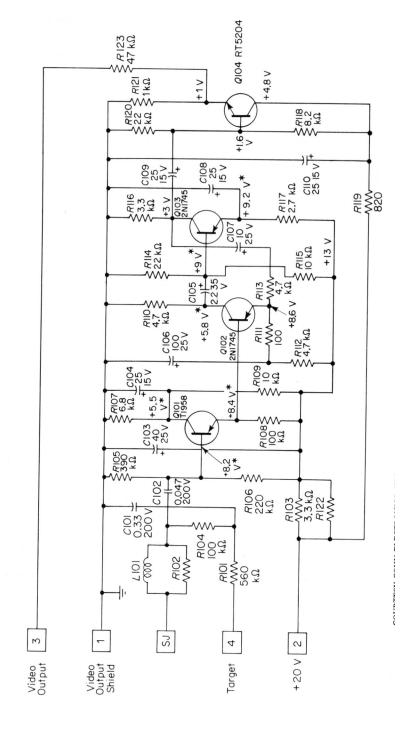

COURTESY COHU ELECTRONICS, INC.

FIGURE 6-2. Another video preamplifier.

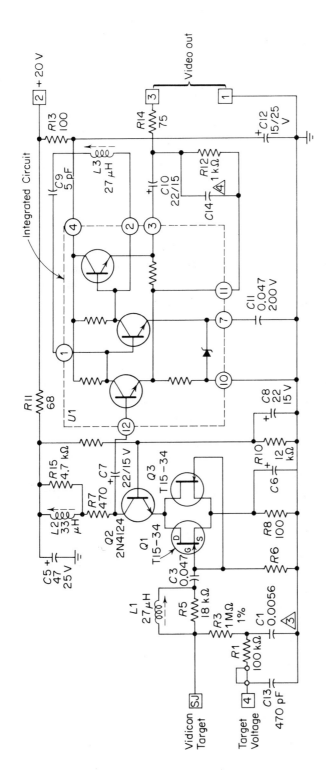

COURTESY COHU ELECTRONICS, INC.

FIGURE 6-3. A video preamplifier using an integrated circuit.

121

$Q1$ and $Q3$ drives the grounded base amplifier, $Q2$. $L2$ is included to provide midband frequency compensation for $Q2$. Output of this circuit is applied to integrated circuit $U1$, which is a three-stage, noninverting feedback amplifier with the necessary low impedance. $L3$ and $C9$ form a trap circuit to compensate for high frequency peaking which occurs in the input circuit. The bandwidth of this circuit exceeds 10 megahertz.

## VIDEO PROCESSING

Video processing amplifiers vary in relative size and complexity according to the applications for which the camera system is designed. Broadcasters, for example, are much more concerned about the quality of their presentation than a user who employs television cameras for simple surveillance. The broadcaster, therefore, uses more elaborate means to

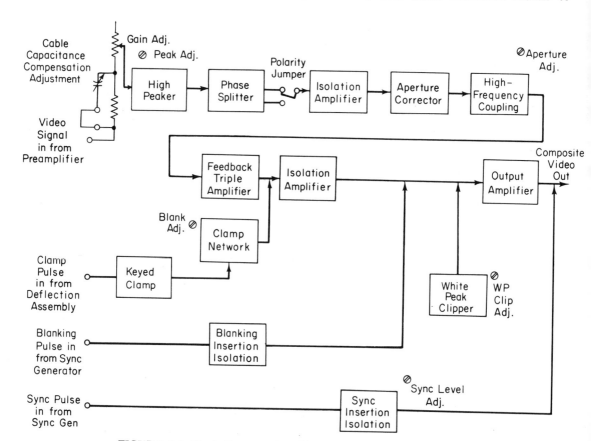

FIGURE 6-4. Block diagram of a video processing amplifier.

condition the signal prior to transmission, and the processing amplifier circuitry is usually quite elaborate. Simple camera systems, on the other hand, may use only a small number of stages to refine the signal.

The order in which the processing functions are performed varies with individual camera systems, and circuit configuration will differ with manufacturers, but the objectives are always the same; remove unwanted elements from the signal, improve picture quality, and provide control (either automatic or manual) over many of the operations that are performed.

Figure 6-4 is a block diagram representation of a video processing amplifier that serves to illustrate the relative position of various circuits in the video flow path. These circuits, and others, are discussed in greater detail in the following pages. The circuits shown are representative of a rather sophisticated video processing amplifier designed primarily for closed circuit television use.

### Video Peaking

It is the purpose of video peaking networks to compensate for the high frequency roll-off in amplitude response that resulted from the high input impedance and distributed capacity of the video preamplifier circuit. These peaking networks may be located in the preamplifier in some instances, as can be noted by reference back to Figure 6-1.

Figure 6-5 illustrates another circuit for achieving video peaking. The transistor provides the desired selective amplification of the incoming

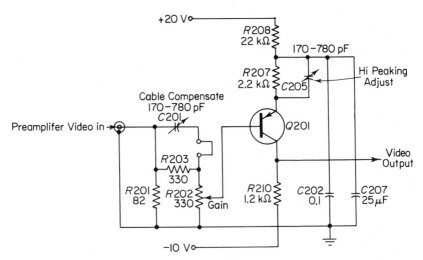

COURTESY COHU ELECTRONICS, INC.

FIGURE 6-5. "Hi-peaking" circuit compensates for high-frequency roll-off.

video. The frequencies affected are controlled by the adjustment of $C205$, which acts as a shunt across a portion of the emitter resistance, providing a low impedance path for the high frequencies and presenting a higher impedance to the lower frequencies. Thus, more degeneration is achieved at the lower frequencies, lowering their relative weight in the output signal.

Excessive "Hi Peaking"

(a)

Insufficient "Hi Peaking"

(b)

FIGURE 6-6. Effects of "Hi-peaking" adjustments.

Video peaking circuits can be thought of as providing control over the square-wave response of an amplifier system. If the video signal were a square wave, too much capacity in the peaking network would result in excessively high frequency response and an *overshoot* would occur during the signal transition (Figure 6-6a). Too little capacity causes slow rise times (Figure 6-6b). The effect of such conditions upon a television presentation is shown in Figure 6-7. Too much compensation causes a signal overshoot as it makes the transition from black to white, yielding a condition known as *trailing whites*. Insufficient compensation causes *trailing blacks,* and proper adjustment results in well defined edges with no trailing effects.

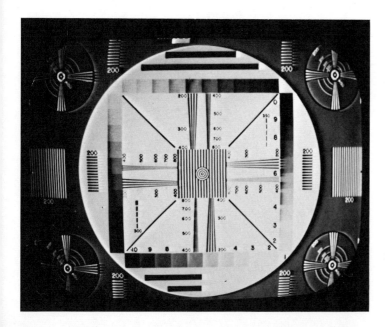

FIGURE 6-7. Excessive "Hi-peaking" produces white smears on transitions from black to lighter shades.

### Aperture Correction

*Aperture correction* is a term associated with the modification of a video signal to compensate for the finite size of the scanning beam spot within the camera pickup tube. Figure 6-8 illustrates the effects of two scanning beam spots moving over a target area which has closely spaced, narrow vertical lines projected upon it. The ideal beam spot is an extremely small point, and the signal it develops is characterized by sharp transitions and good detail contrast. The larger beam spot does not produce as accu-

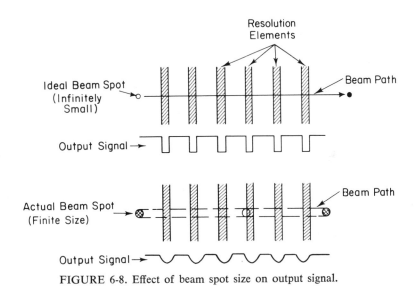

FIGURE 6-8. Effect of beam spot size on output signal.

rate a signal because its size tends to blur the transitions. It may be seen that the larger beam spot will still be discharging some of the white background area even when centered on one of the vertical lines. It would, therefore, provide an integrated output signal which is the combined total of the black and white areas, and the signal would never reach the relative amplitude achieved by the smaller spot. The previous sharp transitions are also impossible to achieve since the integrated signal level changes more slowly as the beam spot progresses into each vertical resolution element.

To compensate for the lower amplitudes and slower signal transitions that are the result of the relatively sizable beam spot, one method is to apply amplitude enhancement to that part of the video signal that constitutes the fine detail portions of the television picture, that is, the high frequency elements. The increased amplitude that results at the higher frequencies is similar to what would occur if the beam spot were made smaller. Although the correction is electronic in nature, the effect is similar to that of causing the beam to pass through an aperture within the camera tube which decreases its cross-sectional area, hence the term *aperture correction*.

Figure 6-9 depicts aperture correction circuitry that uses a miniature transformer to obtain selective enhancement of the high frequency elements

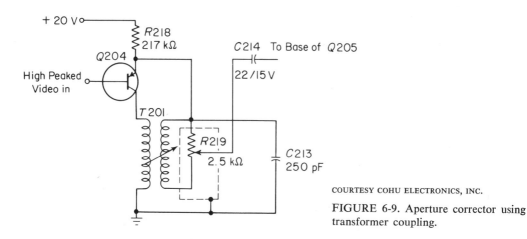

COURTESY COHU ELECTRONICS, INC.

FIGURE 6-9. Aperture corrector using transformer coupling.

of the video signal. When the aperture control potentiometer, *R219*, is positioned so that the arm is connected directly to the emitter of *Q204*, the stage acts strictly as an emitter follower, and couples the signal to the base of *Q205* without any correction being gained. However, when the arm of the potentiometer is advanced, it begins to pick off some of the signal that is being developed across the resistance by the action of the secondary of the aperture transformer. The transformer is frequency selective and couples only the higher frequency elements of the video. It is also wound so that the signal across the aperture potentiometer is of the same polarity as the unenhanced signal, thus its output adds to the original signal, increasing the amplitude of the high frequency elements and leaving the lower frequency elements essentially unaltered.

Figure 6-10 shows another method of achieving aperture correction. This circuit relies upon delay principles rather than selective frequency amplification to achieve enhancement of rapid signal transitions.

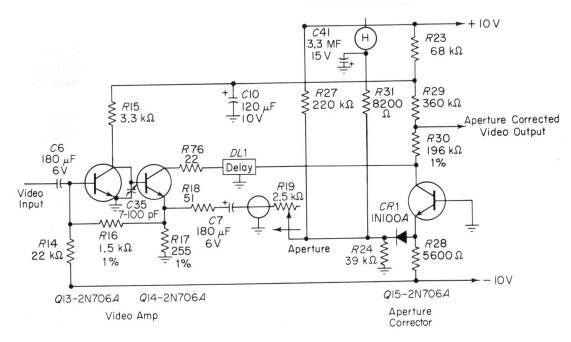

COURTESY RCA.

FIGURE 6-10. Aperture corrector using delay techniques.

Transistors $Q13$ and $Q14$ form a feedback pair amplifier which employs two trimmer capacitors to provide adjustments to control the overall sweep frequency response characteristics of the amplifier. However, the aperture correction function is achieved by taking the output of $Q14$ from both the collector and the emitter. $Q14$ thus acts as a phase splitter, yielding two signals which are identical but opposite in polarity.

The main video path is through the delay line $DL1$, and this signal is delayed for a period of time determined by the characteristics of the delay line. The second signal path from the emitter of $Q14$ supplies an undelayed, reverse-polarity signal to the output. Since this signal is not delayed, it arrives at the output before the main video signal. Thus, if there is a positive-going transition routed through the delay line, a negative-going element will reach the output just before the main signal, causing an *undershoot* of the signal immediately prior to the transition. The undelayed signal at the collector of $Q15$ also travels down the delay line, opposite to the direction of the main video, to the collector of $Q14$. Since the impedance at the collector of $Q14$ is quite high, the signal is reflected back to the collector of $Q15$. Because this signal has traversed the delay line twice, it is delayed by a factor of two and is of reverse polarity. This causes an *overshoot* to occur just after the main signal transition. The delay method of aperture correction therefore causes an enhancing effect both before and after the main video signal transition.

The delay method of aperture correction gives an effect that is somewhat similar to the characteristics of an image-orthicon camera tube. That is, the edging effect is simulated by the delay-type circuit.

### Clamping Circuits

A video *clamping* circuit functions to establish a black reference for the video signal prior to blanking and sync insertion. The vidicon tube normally furnishes a reference voltage during the intervals when blanking pulses are applied for horizontal and vertical retrace. This vidicon output provides a stable dc reference point with which to correlate the black signal level (the signal output which is present when black portions of the scene are being scanned). However, capacitive coupling in the amplifiers ahead of the clamping circuitry do not pass dc components of the signal and the vidicon reference is lost. Clamping of the video signal restores a dc reference and permits proper addition of the blanking and synchronizing signals. Clamping is generally accomplished during the period that will be coincident with the horizontal blanking signal so that the blanking signal, once it is applied, will obscure any disturbing effects that might occur during the actual clamp operation. The video is clamped to its reference dc at the end of each horizontal interval.

Figure 6-11 shows a simple clamping stage that provides keyed clamping action at intervals that coincide with the horizontal scanning frequency. The output from the emitter of transistor $Q305$ is *unclamped* video. However, after the video passes through capacitor $C314$ it is clamped by the action of $L304$ and the two diodes $X301$ and $X302$. The input to diode $X302$ is a positive-going waveform derived from the horizontal scanning section. It is applied to $X302$ through $L305$ and $C315$, turning both diodes on during the positive excursions of the waveform. When the diodes are switched on, ground potential is felt at the video line through $X301$ and $L304$, discharging any other potential that might have been on capacitor $C314$. Since this action occurs after every horizontal line of video, each new line passing through $C314$ has a freshly established dc potential to

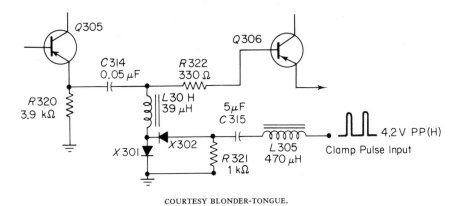

FIGURE 6-11. A keyed clamping circuit.

use as reference. Since the base resistance of $Q306$ is quite high, very little of this potential leaks off during the horizontal line interval and the dc reference remains very stable.

Figure 6-12 illustrates a somewhat different method of achieving clamp action by use of a transistor switch. Here a clamp driving pulse is applied to the base of $Q203$, whose collector is tied directly to the video line. When the transistor is off, as it normally is, video passes undisturbed to the base of $Q204$. During the horizontal retrace interval the clamp drive pulse is applied to the base of $Q203$, turning it on momentarily. During this period, a dc potential, determined by the setting of the *blanking* potentiometer, is applied directly to one side of $C204$, establishing a dc reference potential for the video waveform. In this case, the reference potential is variable, and the potentiometer controls that determine the level are usually front panel controls labeled *Blanking Level,* or *Pedestal.*

### Blanking Insertion

Blanking is inserted onto the video signal to assure proper beam cutoff in the television monitor or receiver. Since the blanking waveform is impressed onto the video at intervals corresponding to vidicon retrace, it also eliminates transients and extraneous noise that might have been generated during camera retrace.

Blanking of the video signal is often accomplished by merely shorting the video line to ground. Figure 6-13 illustrates such a method. The in-

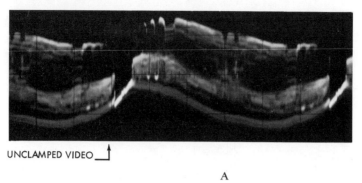

UNCLAMPED VIDEO

A

Clamp In

R201
150 Ω
−10 V

C204
0.22/35 V

+7 V

Video In

Q202
2N2400

+14.1 V

R208
2200 Ω

C205
15/25 V

R209
56 kΩ

C206
22/15 V

R210
2.5 kΩ
CW

Blanking

Q203
2N2188

R212
1000 Ω

2N2188
Q204

Clamped Video Out

R213
10 kΩ

C207
22/15 V

RT 201
T

R207
1500 Ω

R211
2700 Ω

+18 V

B

CLAMPED VIDEO

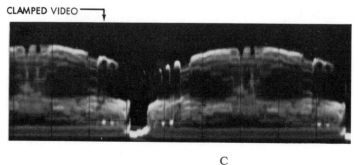

C

COURTESY COHU ELECTRONICS, INC.

FIGURE 6-12. Clamp circuits eliminate low-frequency hum.

coming blanking signal (that was generated in a sync generator or other circuitry) is applied to the base of $Q209$, a transistor switch which is normally biased off. The input pulse turns the transistor on and shorts the video line directly to ground potential, thereby eliminating any video signal during that interval.

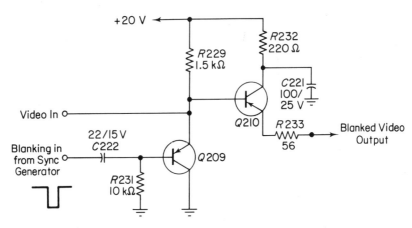

COURTESY COHU ELECTRONICS, INC.

FIGURE 6-13. Video blanking is accomplished by switching Q209.

### Sync Insertion

Since sync pulses must be negative-going with respect to the blanking signal level, an element of the blanked area of the video signal is simply driven to a more negative potential. This is generally accomplished as one of the last operations performed to the video signal within a camera system.

Figure 6-14 illustrates one means of accomplishing sync insertion, which gives control over the amplitude of the sync pulses once they are applied to the video. Sync pulse amplitude on the composite video waveform is normally set to 0.4 volt for closed circuit application and 0.3 volt for broadcast use (assuming the system employs an EIA sync generator, see Chapter Seven).

Some industrial television cameras do not employ sync signals, as such, but simply rely upon negative blanking pulses of rather high amplitude to serve as sync also.

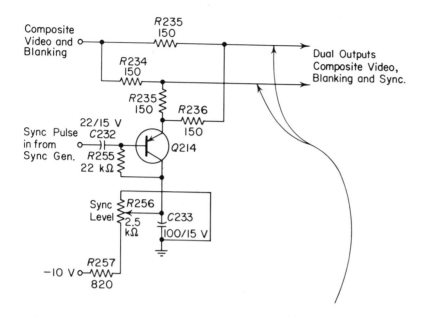

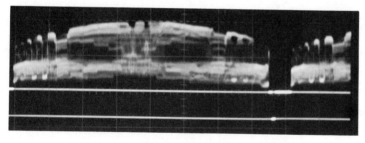

FIGURE 6-14. Sync addition completes the composite monochrome video waveform.

### Polarity Reversal

Many times it is a desirable feature for a camera system to be able to reverse the polarity of the video signal. This is sometimes done for special effects, but it is especially desirable in a television camera that is to be used as a film-chain system. By reversing the polarity of the video, it is then possible to view motion picture or still negatives instead of having to process them into positives. In broadcast television this can be a great time saver in the airing of newsfilm.

By use of a phase-splitter stage and a relay, polarity reversal can be

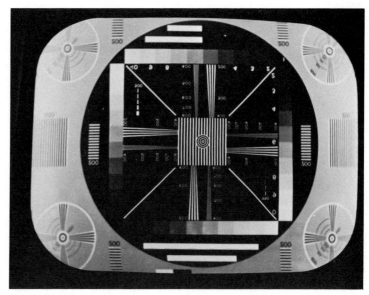

FIGURE 6-15. Polarity reversal yields a negative image.

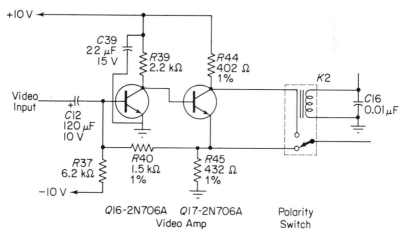

FIGURE 6-16. Polarity reversal circuit uses a relay.

quite simply accomplished (Figure 6-16). The video at the emitter of the transistor is of the same polarity as that of the input and the signal at the collector is inverted. Proper selection of resistance values assures that both signals are of the same amplitude, though opposite in polarity. It then becomes simply a matter of choosing the necessary polarity by properly positioning the relay. It should be noted that the relay is shown as being shielded. Such precautions are common, reducing the possibility of noise and oscillation.

It is, of course, necessary to accomplish polarity switching before blanking or sync are added to the video waveform, since these must maintain a negative-going characteristic regardless of the video polarity.

### White Peak Clipping

White peak clipping is sometimes used in cameras to assure that the positive excursion of the video signal will not exceed a preset level. Any signal that would surpass this level is simply clipped off, so the device is generally used only in extreme cases. The clipper finds use where the contrast between low and highlight areas in the same scene is too great to be represented on the monitor. The clipper can also be used as a safeguard to prevent unexpected signals above a certain amplitude from upsetting equipment such as function generators, tape recorders, etc., which utilize the video output.

In Figure 6-17, emitter follower $Q8$ has a diode at its base that couples video signal to ground when conducting. Since the video output at the emitter is positive-going, all video would be routed through capacitor $C40$ to ground, if it were not for a positive dc voltage placed at the cathode of

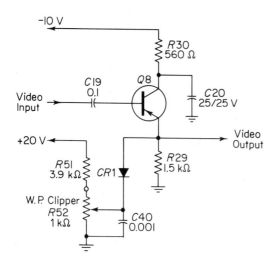

COURTESY COHU ELECTRONICS, INC.

FIGURE 6-17. White-peak clipping eliminates high-amplitude signal excursions.

*CR*1. The value of the dc voltage will determine at what positive potential of the video waveform the diode will conduct, and is controlled by the setting of the *White Peak Clipper* control, *R*52.

As previously suggested, the clipper should be used only in special circumstances. Normally the diode should be reverse biased to the extent that even a signal of extreme amplitude will be fully represented on the monitor. Thus, extremely bright objects will be immediately apparent on the monitor and the camera can be moved, or the lens capped, before any damage to the camera tube can result.

## Gamma Correction

The term *gamma* has had various meanings ascribed to it, depending upon the discipline in which it is used. In television, gamma characteristics are primarily related to the luminance characteristics of the television camera tube, the monitor picture tube, and the linearity of the circuits between the two.

Most cathode ray display tubes do not maintain a linear relationship between the driving voltage and the luminance output. The various luminance elements in a monitor presentation are therefore distorted with respect to the voltage values of the input signal. In other words, increasing the driving voltage by a factor of two would not necessarily increase the luminance by the same factor.

Camera tubes also possess gamma characteristics that must be considered when determining the linearity with which different values of luminance are displayed. Most camera tubes do not provide an output that is linear in amplitude with respect to the intensity of the scene brightness. This was noted in Chapter Four when light transfer characteristics were discussed. The light transfer characteristic, when plotted on log-log paper, generally approximates a straight line. One of the primary reasons for the use of log plotting is that the eye's response to light also approximates a logarithmic function.

Once the light transfer function is plotted (Figure 6-18), its slope is measured. If the line passes through the origin, the system is linear and is said to have a gamma of 1.0. If it deviates above or below this, measuring the slope yields the effective gamma of the system. Thus, gamma is a measure of the *slope* of the light transfer characteristic.

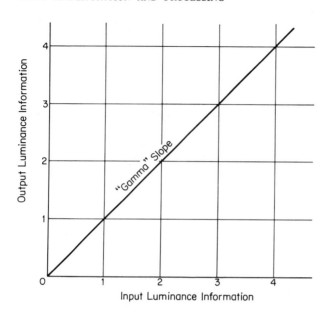

FIGURE 6-18. When input and output luminance information are equal, a gamma of 1 is achieved.

From the above, it can be appreciated that any differential gain irregularities in the amplifying or transmission equipment in a video system will also affect the gamma of the final presentation. In fact, selective control of differential gain (gain vs. amplitude) in an amplifier is often used to compensate for the nonlinear characteristics of a camera tube or kinescope.

Vidicon television systems that are used for live pickup generally do not require such measures because the gamma characteristics of the vidicon and the kinescope tend to offset one another, resulting in an overall characteristic of approxmiate unity. However, image-orthicon and other tubes have entirely different gamma curves, and correction circuits are generally incorporated in the camera to compensate. When television cameras are used to view photographic film, it is also necessary to introduce correction to compensate for the gamma characteristics of the film.

Figure 6-19 illustrates a stair-step signal that is used as a convenient means to measure the gamma characteristics of an amplifier system. Figure 6-19a is a linear stair-step. However, if it were fed into a monitor and viewed on the kinescope it would not appear linear. That is, the different shades of grey that each voltage level creates would not appear to be progressively equal in contrast change. On the other hand, the signal in Figure 6-19b would present a monitor display that would appear linear because

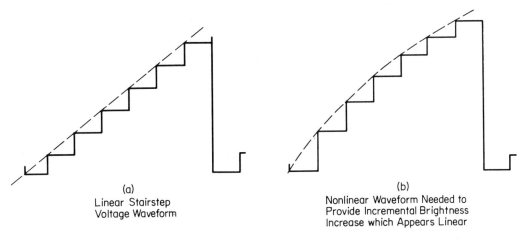

(a)
Linear Stairstep
Voltage Waveform

(b)
Nonlinear Waveform Needed to
Provide Incremental Brightness
Increase which Appears Linear

FIGURE 6-19. Effects of amplifier gamma characteristics on a stairstep waveform.

the distortion that appears in the signal compensates for the nonlinear
characteristics of the picture tube output.

Figure 6-20 illustrates one method of providing for gamma control
in a video amplifier or processor. $R1$, $R4$, $CR1$, $R7$, $R10$, $CR4$, and $R13$
form the gamma gain-modifying components. When the video level is low,
both of the diodes are biased "on." The impedance felt in the emitter of
the transistor stage is essentially the parallel values of $R13$, $R7$, and $R1$,

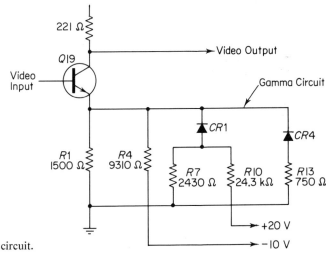

FIGURE 6-20. Gamma circuit.

in addition to the effect of the biasing resistors $R2$ and $R10$. The resistance that is present controls the amount of degeneration in the transistor amplifier. As the video level increases, $CR4$ becomes biased "off," and the total amount of resistance is increased because of the loss of one parallel element. If the video input rises still further, $CR1$ becomes biased "off," leaving only the impedance of $R1$ and $R4$ to consider. Since another parallel path has been removed, impedance is still higher. Because the higher impedance causes more degeneration in the transistor stage, a maximum video signal input produces minimum gain. Such impedance changes are chosen so that the change in video gain in the amplifier closely approximates the gamma characteristics needed. The circuit in this example operates at a fixed value of gamma correction.

### Automatic Target Control

Automatic control over the target voltage in vidicon cameras causes the camera to vary its sensitivity in relation to the amount of scene illumination. Cameras thus equipped can operate over extremely wide dynamic light ranges without manually adjusting the lens or other electrical controls.

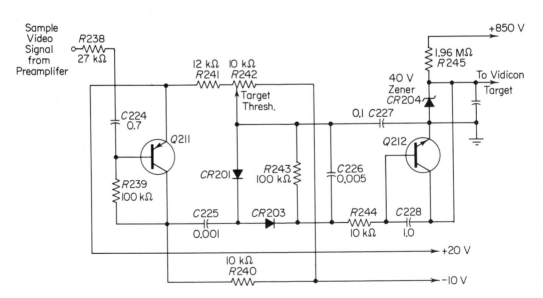

COURTESY COHU ELECTRONICS, INC.

FIGURE 6-21. Automatic target control circuit samples video waveform.

In the example of Figure 6-21, a sample of the video signal from a preamplifier output is applied to a detector circuit. Video from the preamplifier is amplified by $Q211$ and applied to a voltage doubler ($CR201$, $CR203$, $C225$, and $C226$), producing a dc voltage that is proportional to the video amplitude at the input. The target threshold voltage, controlled by potentiometer $R242$, adjusts the relative sensitivity of the circuit by applying a bias voltage to which the video-derived dc voltage is added, and applied to the base of $Q212$. As the video signal increases at the input (signifying an increase in scene illumination), the dc voltage at the base of $Q212$ goes more positive, causing the amplifier to conduct harder and placing its collector nearer to ground potential. Since the collector is connected to the vidicon target, the increased video input has resulted in a decrease in target voltage. The target voltage decreases until the steady-state condition of the circuit is re-established and the resultant video at the input achieves its original amplitude. Should the video input decrease in amplitude, the collector voltage of $Q212$ would rise, and would thus increase the sensitivity of the vidicon and provide more video signal to the circuit input.

### RF Modulators

Many times it is desirable to have a television camera system that will operate directly into a standard home television receiver. In such instances it is usually not practical to modify the receiver to accept nonmodulated video, so an RF modulator is added to the camera that will operate on one of several standard VHF television channels (usually 1 through 6).

The circuit in Figure 6-22 is typical of the methods used for RF modulators in cameras designed for industrial and home use. $A501$ is an RF oscillator that operates at a frequency determined by the setting of the slug-tuned coil in its collector circuit. $Q502$ serves as a mixer stage, modulating the RF signal with the video waveform applied to its emitter. The modulated RF signal at the collector of $Q502$ is applied at the output through $C508$.

Since most cameras of this type utilize a 75-ohm coaxial cable to route video signals, the input to the receiver will generally require an impedance-matching transformer to allow the best signal transfer into the 300-ohm receiver input.

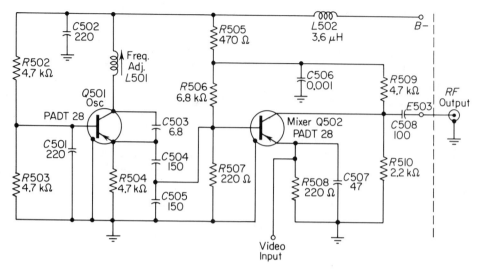

FIGURE 6-22. RF modulator allows video waveform to be displayed on standard television receiver.

### An 8-Megahertz Amplifier

Figure 6-23 on pages 142 and 143 illustrates a complete video amplifier system (less output stage) designed for use in closed circuit applications. While it does not aspire to broadcast uses, it is certainly adequate for a great number of industrial and educational applications. Amplifier circuits such as this are employed in a majority of television systems used for educational, surveillance, and light industrial tasks.

The four-stage amplifier consisting of $Q301$ through $Q304$ employs tunable inductors for control of video bandwidth and peaking. Gain is effected by controlling the degeneration at the emitter of $Q303$ with potentiometer $R310$. A composite waveform containing both horizontal and vertical pulses is applied to the cathode as blanking information, and is routed to the emitter of $Q304$ where it is added to the video signal as a form of "industrial" sync. The amplitude is made adjustable by potentiometer $R319$. The driven clamp circuit appears at the junction of $C314$ and $R322$, using switching diodes energized by the positive-going signal (occurring at the horizontal rate) that appears at the anode of diode $X302$.

Aperture correction is achieved by the action of transformer $T301$ which enhances the amplitude of the higher frequencies to an extent determined by the setting of the aperture control, potentiometer $R324$. The video is then routed to the output through emitter follower $Q306$. The video at this point is negative-going (inverted) and must rely upon an output stage (not shown) to invert it and drive the video cable to the monitor.

COURTESY BLONDER-TONGUE.

FIGURE 6-23. An 8-MHz amplifier.

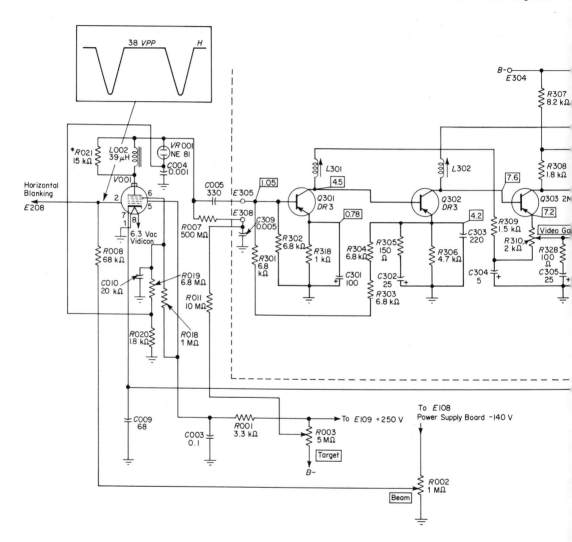

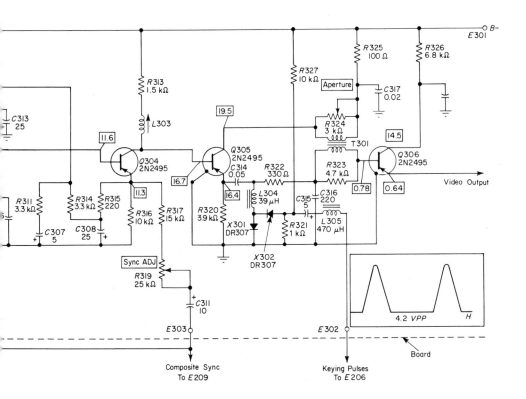

4.2 VPP    H

Composite Sync
To E 209

Keying Pulses
To E 206

# 7

# MONOCHROME SYNCHRONIZING GENERATORS

The scanning system in a monitor or receiver must reproduce exactly the sequence of events that takes place in the camera scanning system. To accomplish this, it is necessary that the two be synchronized in a precise manner. Horizontal and vertical scan and retrace in both the camera and monitor must occur at identical intervals with respect to the video information to obtain a useful presentation. Anyone who has ever misadjusted the *vertical-hold* or *horizontal-hold* control on a home television receiver is familiar with the results obtained when a monitor is not synchronized or locked to the camera system.

Another important function of a sync generator is to maintain a rigid phase relationship between the horizontal and vertical scanning systems to assure the stable 2:1 interlace demanded by many applications. To achieve this, countdown circuitry is generally incorporated to provide a vertical drive signal that is an exact submultiple of the horizontal frequency.

It should be understood that a sync generator is not generally considered a part of the deflection circuitry. It simply serves as a source of accurate timing signals which are used to trigger or drive the deflection circuits in both the camera and the monitor. In more sophisticated systems it will also provide horizontal and vertical blanking signals to be inserted onto the video waveform.

COURTESY BLONDER-TONGUE.

FIGURE 7-1. A random interlace deflection generator.

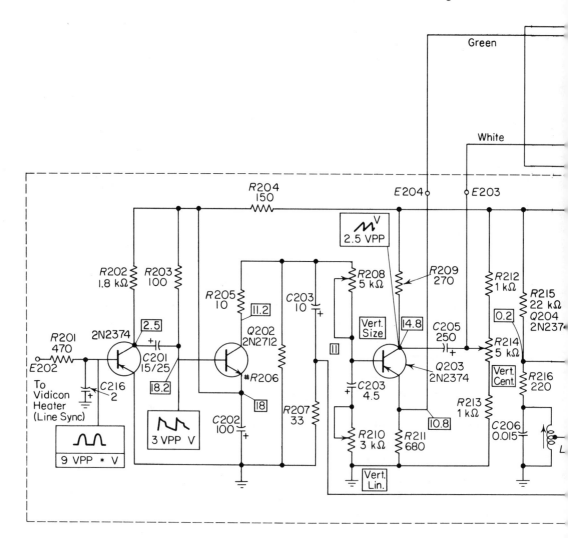

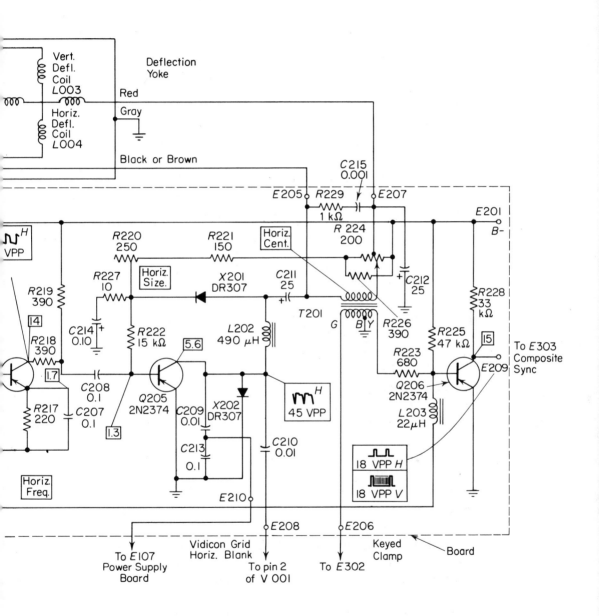

## PRINCIPLES OF SYNCHRONIZATION

### Industrial Sync

In the earlier days of television, it became apparent that TV camera systems could play a very useful role in many areas of industry. However, broadcast-type systems were generally quite bulky, consisting of large racks of equipment, and cost was an important prohibitive factor. To fill the need, systems were developed specifically for the industrial user. In such equipment, circuitry was generally reduced to a minimum to meet the demands of reduced cost and limited space. The performance of these systems was not acceptable for most broadcasting purposes but was considered adequate to the demands of the industrial market. One important difference was the less stringent requirements for the synchronizing waveforms utilized, hence the term *industrial sync*.

While the term industrial sync normally refers to the actual configuration of the synchronizing waveform as added to the video signal, common usage has made the definition somewhat broader in scope. For instance, many television systems do not utilize a sync generator as such, but are commonly included in the same general category. In such cases, deflection signals for the horizontal and vertical are derived independently and are not locked to each other in a common phase relationship. Thus, the positioning of the horizontal lines of each succeeding vertical field is not fixed, and the line spacing may vary in a completely random nature. This gives rise to the expression *random interlace*. This method of scanning is used in many less costly television cameras. In these systems the vertical oscillator in the camera is generally triggered by the 60-hertz power line frequency, while the horizontal oscillator is either crystal controlled or of the free-running type that is governed only by the time constants of the components within its circuitry.

The schematic of Figure 7-1 illustrates an example of horizontal and vertical deflection circuitry that operates to produce a random interlace raster. The vertical yoke is driven by a sawtooth current waveform that is derived directly from a 60-hertz input through $Q201$, $Q202$, and $Q203$. The horizontal deflection transistor, $Q205$, obtains its drive from the free-running horizontal oscillator, $Q204$. It may be noted that there is no means incorporated to maintain a constant phase between the horizontal and vertical deflection systems. $Q206$ combines samples of the horizontal and vertical waveforms and routes them to the video amplifier where they will be added to the video signal as composite sync.

Industrial television systems may typically employ the blanking excursions of the video signal to trigger the deflection circuitry within the monitor. Figure 7-2 shows such a signal and the area that is used for the triggering. Since the blanking pulses for the horizontal retrace periods are quite narrow with respect to the blanking pulse necessary for the vertical retrace (due to the large difference in retrace times), the two may be separated quite easily by special circuits in the monitor. This eliminates the need for routing each to the monitor by its own separate path. Such a method of synchronizing the monitor depends upon the blanking pulses being rather high in amplitude and extending below the level of the video waveform to enable the monitor circuitry to distinguish them.

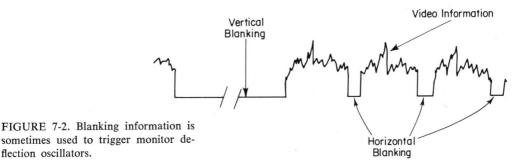

FIGURE 7-2. Blanking information is sometimes used to trigger monitor deflection oscillators.

Industrial sync is sometimes found wanting in its ability to maintain a stable 2:1 interlace within a monitor, even when countdown circuitry is used to achieve the proper phase conditions. At times, the horizontal lines of successive vertical fields will not fall exactly between the lines of the previous field, even though the phase relationship between the horizontal and vertical blanking excursions remains constant. It is even possible for the horizontal lines of succeeding vertical fields to be superimposed upon one another, particularly if the *vertical-hold* control on the monitor is slightly misadjusted. This effect is known as *line pairing,* and when it occurs the vertical resolution capabilities of the system diminish markedly. This is obvious when it is considered that the number of horizontal scanning lines is cut in half when a monitor is pairing.

Another difficulty that may arise in connection with some types of industrial sync can be attributed to a lack of horizontal sync information during the vertical sync period. If the horizontal oscillator in the monitor has a free-running frequency somewhat different from that of the horizontal sync frequency, the oscillator will revert to its free-running state during the period of vertical sync. Thus, after the period of vertical sync, the horizontal sync pulses must pull the horizontal oscillator of the monitor back onto frequency. This can cause a curved type of picture offset at the top of the raster, often referred to as *monitor pulling*. It can also cause a horizontal jitter at the extreme top of the picture, known as *flag waving*.

### Broadcast Television Synchronizing Standards

Broadcast television and many closed circuit television applications require a method of synchronization considerably more sophisticated than that offered by industrial synchronizing methods. A stable 2:1 interlace is demanded, and a very precise method of driving the monitor circuits is necessary. Accurate blanking information must also be generated to assure that retrace occurs unobserved.

Figure 7-3 shows one horizontal line of video information with the horizontal blanking and sync waveforms inserted. These waveforms are originated within the sync generator and are impressed onto the video signal during processing. It should be noted that the black information within the video signal is positioned just above the *blanking level*. The blanking level represents the absolute black that is to be displayed by the monitor, and it will effectively cut off the electron beam within the picture tube of a properly adjusted monitor or receiver. The blanked area of the video signal is coincident with the blanking that takes place in the camera, but it is inserted onto the video in one of the later stages of processing and is not to be confused with the camera blanking provided for the pickup tube.

The negative-going pulse that occurs during the horizontal blanking interval is the *horizontal sync* pulse, and it is this pulse that is used to trigger horizontal retrace in the monitor, keeping it in sync with the camera horizontal retrace. Horizontal scanning in the monitor is, therefore, made coincident to that of the camera and retrace is caused to occur during the time that the blanking pulse is present.

Figure 7-4 shows an expanded representation of the horizontal blanking interval on a composite video waveform. The video that appears just prior to blanking would be displayed on the monitor at the right side of the screen. The area of the blanking pulse that occurs before the sync pulse

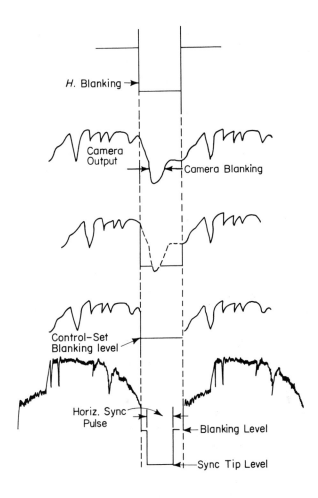

FIGURE 7-3. Horizontal blanking and sync waveforms as inserted onto video signal.

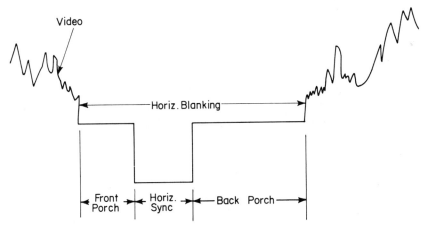

FIGURE 7-4. The horizontal blanking interval.

151

interval is termed the *front porch*. Blanking is initiated prior to the horizontal sync interval to assure beam cutoff before retrace is started. The portion of the blanking interval that occurs after the sync pulse is called the *back porch*. The extended period of blanking following the horizontal sync pulse is necessary to allow the monitor to complete retrace and begin a normal scan line before bringing the beam out of cutoff. Blanking pulse duration is also important in establishing the 4:3 aspect ratio of the active video display.

Figure 7-5 illustrates the synchronizing signal as it appears during the vertical retrace portion of the scanning interval. This waveform depicts the vertical blanking portion of the composite video waveform and displays the various pulses present. The vertical sync pulse interval is distinguished by the comparatively wide negative-going pulses. It is this vertical pulse interval that initiates the vertical retrace in the monitor. This portion of the signal has a more negative dc average than that which is contained within the rest of the vertical deflection period, and there is circuitry incorporated in receivers and monitors which senses this and causes vertical retrace to be initiated. One might well wonder why a series of wide pulses was used instead of a single negative pulse which would occupy the same time interval. Attention to the waveform will show that the vertical sync pulse interval is indeed the large negative pulse mentioned, but it is interrupted or "serrated" by narrow positive excursions which are occurring at twice the horizontal frequency. These pulses are called *serration pulses*.

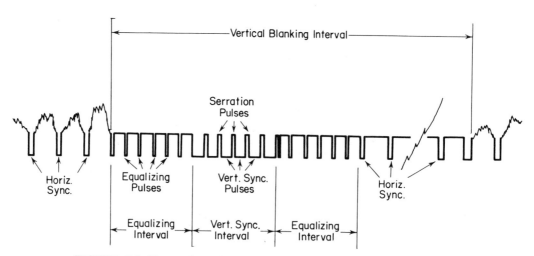

FIGURE 7-5. The vertical blanking interval.

It can also be noted that immediately prior to the vertical interval, and just after it, occur a series of six narrow negative pulses of the same frequency as the serration pulses. These pulses are called *equalizing pulses* and, when used in conjunction with the serration pulses, serve to assure that the horizontal frequency of the monitor remains locked and completely stable during the vertical retrace period. As was pointed out in the discussion of industrial sync, an absence of pulses during this time would allow the horizontal oscillator to drift out of phase with the horizontal frequency of the camera. Because it would have to be pulled back into phase by the sync pulses after each vertical interval, the possibility exists that a few lines of video information at the top of the raster would be disrupted.

The fact that the frequency of the equalizing and serration pulses is double that of the horizontal frequency is necessitated by the 2:1 interlace that is employed. Figure 7-6 illustrates the relationship of two successive vertical blanking intervals. It may be seen that the first waveform begins vertical blanking coincident with horizontal blanking, ending the vertical scan with a full horizontal line of video information. The second waveform illustrates the one-half horizontal line of video that ends the following vertical interval, demonstrating that an interlacing of the horizontal lines

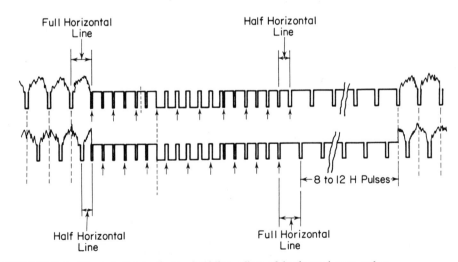

FIGURE 7-6. Alternate fields display half-line offset of horizontal sync pulses with respect to the vertical interval. Arrows indicate vertical interval pulses that are accepted by the monitor as triggering.

does indeed occur. It may also be noted that the start of each vertical field begins alternately with a half line and a whole line.

The arrows below the pulses in Figure 7-6 indicate those used to trigger the horizontal oscillator in the monitor. Alternate equalizing and serration pulses are used during each alternate vertical period. The time constant of the horizontal oscillator in the monitor is such that it will be triggered only by every other pulse when they occur at twice the horizontal rate. It accepts such triggering as a substitute for the horizontal sync pulses and remains phase-locked during the vertical retrace period.

**Sync Utilization**

In order to understand more thoroughly the broadcast type sync waveform, it is helpful to examine briefly how it is utilized in the monitor or receiver.

Figure 7-7 schematically illustrates a method of separating the horizontal sync information from the overall sync waveform. Transistor $Q1$ is an amplifier that has a differentiating network in its base circuit. This

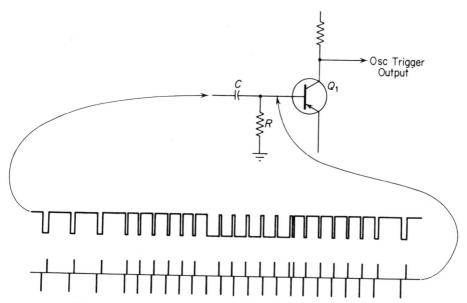

FIGURE 7-7. Differentiating circuit extracts horizontal information.

transforms the sync pulses into a series of sharp spike-like positive and negative pulses, the duration of which is determined by the amplitude of the input pulse and the values of $R$ and $C$. In this case, the negative-going excursions are used by the amplifier to develop the drive pulses for the oscillator. Note that the equalizing and serration pulses also produce outputs from this network which are very similar in nature. The time intervals between these pulses are constant full-line or half-line increments, assuring that the oscillator has a very accurate time base for good frequency stability.

In Figure 7-8, transistor $Q1$ is an amplifier that has an integrating network in its base circuit. Again, the values of $R$ and $C$ and the input pulse amplitude determine the output from the stage. The values are chosen such that the capacitor will be able to build a charge sufficient to trigger the vertical oscillator only if a pulse of long duration is applied. It can be noted that short interruptions in this pulse, such as that caused by the serration pulses, will not be sufficient to discharge the capacitor appreciably, and serve only to make the charging curve somewhat irregular. The horizontal sync pulses do not allow the capacitor to charge to any great extent due to the long discharge interval between the pulses. This is also true of the equalizing pulses which, although they occur at twice the frequency, are of narrower width than the horizontal sync pulses and therefore maintain a similar dc average.

The integrating circuit can be thought of as a device that is sensitive to the average dc level of a pulse train, and since the vertical sync pulse interval has an average dc level that is much more negative than any of the other pulses, it is this effect that will ultimately cause an output from the stage and will serve to trigger the vertical oscillator.

It is the purpose of the equalizing pulses to assure that vertical retrace occurs at precisely the same time in each vertical interval, thus establishing a positive 2:1 interlace. Since each alternate vertical field ends with a half-line difference, one would expect the integrating circuit to begin its vertical sync charge time at a slightly different potential for each alternate field, due to the charge left by the last horizontal sync pulse. If it were not for the effect of the equalizing pulses upon the integrator, this could certainly be the case. However, as can be seen from the diagram (Figure 7-8), the equalizing pulses do indeed equalize the potential at which the capacitor begins its charge. Consequently, the vertical oscillator is triggered at the same critical point during the vertical sync period, guaranteeing a stable 2:1 interlace.

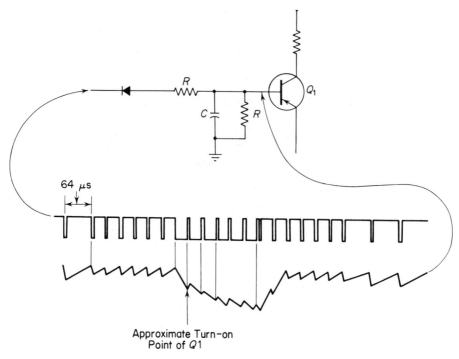

FIGURE 7-8. Integrating circuit develops vertical trigger.

## STANDARD SYNC
## GENERATOR WAVEFORMS

Most sync generators used by broadcasters in the United States, as well as a great many of those used in closed circuit television systems, provide output waveforms that conform to the standards of the Electronic Industries Association (EIA). This serves to eliminate misunderstandings between manufacturers and users, and allows interchangeability and versatility among various pieces of equipment of different manufacture.

There are presently three important standards of the EIA that pertain to the configuration of the synchronizing waveform in television systems. These standards list such important parameters as pulse widths, amplitude, rise times, phase relationship, etc. The three signal configurations recommended in these standards are as follows:

EIA standard RS-170, *Electrical Performance Standards for Monochrome Television Studio Facilities*, recommends synchronizing waveforms which conform to the illustrations in Figure 7-9.

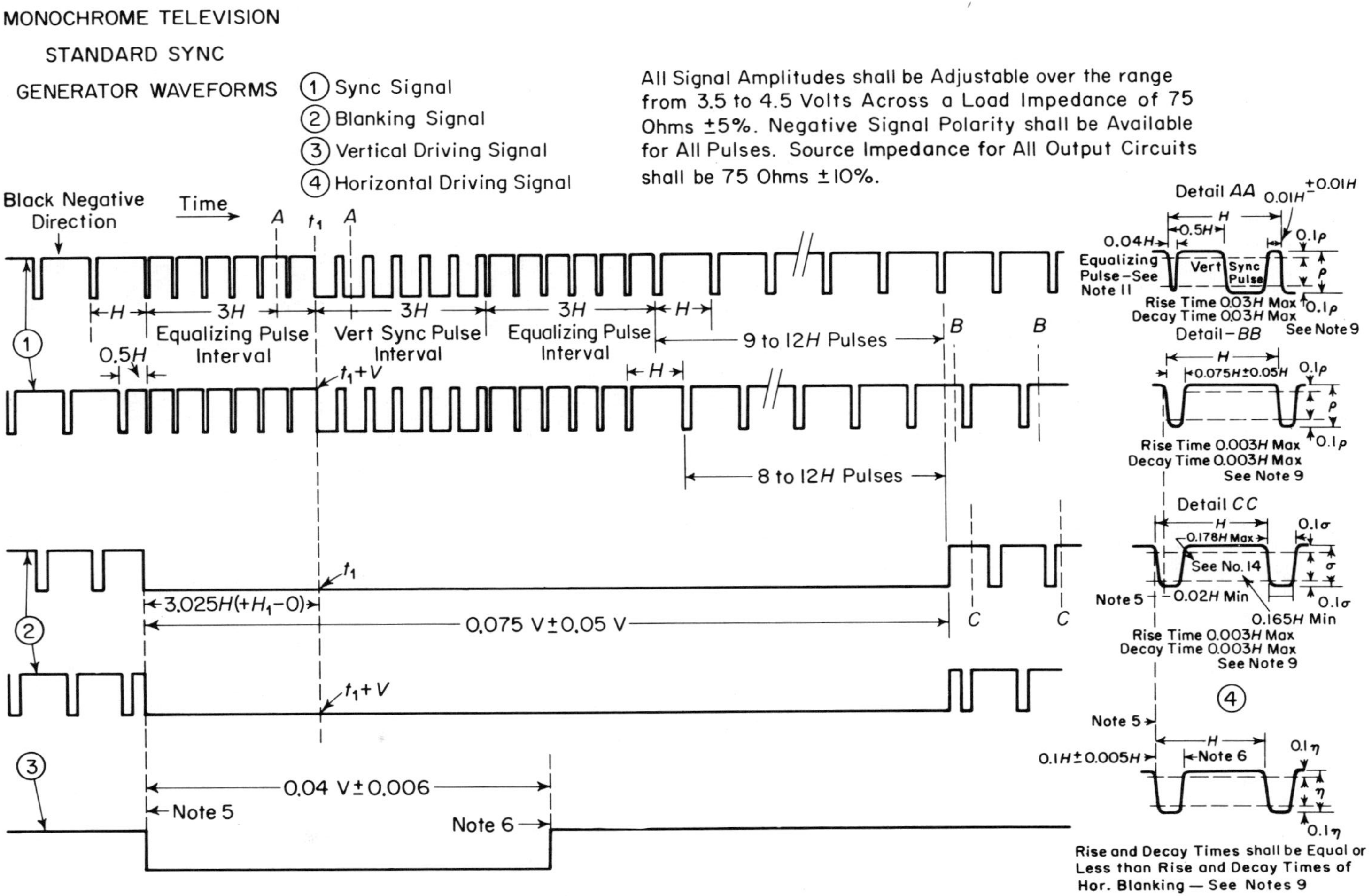
MONOCHROME TELEVISION
STANDARD SYNC
GENERATOR WAVEFORMS
(1) Sync Signal
(2) Blanking Signal
(3) Vertical Driving Signal
(4) Horizontal Driving Signal
All Signal Amplitudes shall be Adjustable over the range from 3.5 to 4.5 Volts Across a Load Impedance of 75 Ohms ±5%. Negative Signal Polarity shall be Available for All Pulses. Source Impedance for All Output Circuits shall be 75 Ohms ±10%.
Black Negative Direction
Time
Equalizing Pulse Interval
Vert Sync Pulse Interval
Equalizing Pulse Interval
9 to 12H Pulses
8 to 12H Pulses
3.025H(+H1−0)
0.075 V±0.05 V
0.04 V±0.006
Note 5
Note 6
Detail AA
Equalizing Pulse–See Note 11
Vert Sync Pulse
Rise Time 0.03H Max
Decay Time 0.03H Max
See Note 9
Detail–BB
Rise Time 0.003H Max
Decay Time 0.003H Max
See Note 9
Detail CC
0.178H Max
See No. 14
0.02H Min
0.165H Min
Rise Time 0.003H Max
Decay Time 0.003H Max
See Note 9
0.1H±0.005H
Rise and Decay Times shall be Equal or Less than Rise and Decay Times of Hor. Blanking — See Notes 9

COURTESY EIA.

FIGURE 7-9. Standard broadcast sync generator waveforms.

To conform to the recommendations of RS-170, monochrome sync generators are required to provide four separate and distinct output waveforms: (1) sync signal, (2) composite blanking signal, (3) vertical driving signal, (4) horizontal driving signal. The sync and blanking signals are added to the video for use at the monitor or receiver, and the vertical and horizontal drive signals are used to trigger the respective deflection circuits in the camera system or systems. A strict phase relationship is maintained between all of the waveforms. Sync and blanking waveforms are shown for two successive fields to illustrate the offset of horizontal sync that assures interlace.

It may be noted that all pulse durations and time intervals are specified with respect to the time it takes to complete one horizontal ($H$) or vertical ($V$) period. Thus, in a 525-line system, one $H$ would equal about 63.5 microseconds and one $V$ would equal 1/60th of a second or about 16.67 milliseconds.

This same standard is often used to specify synchronizing waveforms for television systems which utilize higher scanning ratios, but the standard was written for 525/60 scanning standards only. For a synchronizing system to meet this standard, each and every requirement shown on the diagram must be fulfilled.

Figure 7-10 illustrates the configuration which the synchronizing signals, as added to the composite video signal, must take to conform to EIA standard RS-330, *Electrical Performance Standards for Closed Circuit Television Camera 525/60 Interlaced 2:1*. This standard applies only to 525-line closed circuit systems, but it does not in any way indicate that these same systems may not use synchronizing waveforms that would exceed these standards, as would those of RS-170.

It should be noted that RS-330 does not require the use of equalizing or serration pulses during the vertical interval. However, if they are used, they must conform to the requirements shown. The relationship between horizontal sync and horizontal blanking is rigidly established. Since the pulse durations and rise times are specified here in microseconds, this standard cannot be used even as a reference to specify waveform configuration for closed circuit systems of higher scan rate.

Figure 7-11 illustrates the recommended waveforms obtained from EIA standard RS-343, *Electrical Performance Standards for High Resolution Monochrome Closed Circuit Television Camera*. They are basically the same as those shown in Figure 7-10, but these are for use with scanning systems that employ scan ratios of 675, 729, 875, 945, 1023, etc., lines-

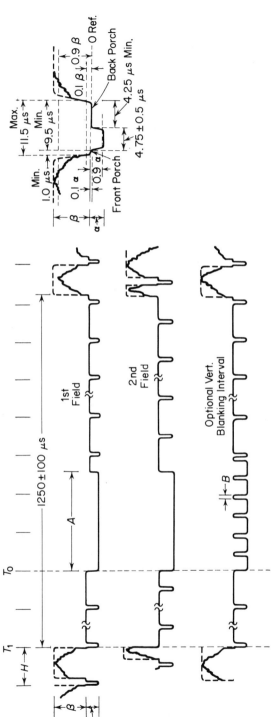

Notes:

1. $\beta = 0.714 \pm 0.1$ Volt (100 IRE Units).

2. $\alpha = 0.286$ (40 IRE Units) Nominal.

3. Sync to Total Signal Ratio $\left(\dfrac{\alpha}{\beta + \alpha}\right) = (28.6 \pm 5)\%$.

4. Blanking = $7.5 \pm 5$ IRE Units (2.5% to 12.5% of $\beta$).

5. Horizontal Rise Times Measured from 10% to 90% Amplitudes shall be Less than 0.3 $\mu$/s.

6. Overshoot on Horizontal Blanking Signal shall not Exceed 0.02 $\beta$ at Beginning of Front Porch and 0.05 $\beta$ at End of Back Porch.

7. Overshoot on Sync Signal shall not Exceed 0.05 $\beta$.

8. $T_0$ = Start of Vertical Sync Pulse.

9. $T_1$ = Start of Vertical Blanking.

10. $T_1 = T_0 \pm 0.250$ $\mu$/s.

11. $A$ – Vertical Sync Pulse = $150 \pm 50$ $\mu$/s Measured Between 90% Amplitude Points.

12. Rise and Fall Time of Vertical Blanking and Vertical Sync Pulse, Measured from 10% to 90% Amplitudes shall be Less than 5 $\mu$/s.

13. Tilt on Vertical Sync Pulse shall be Less than 0.1 $\alpha$.

14. If Horizontal Information is Provided During the Vertical Sync Pulse it Must be at 2H Rate and as Shown in the Optional Vertical Blanking Interval Waveform.

15. $B$ – Vertical Serration = $4.5 \pm 0.5$ $\mu$/s Measured Between the 90% Amplitude Points. Rise Times Measured from 10% to 90% Amplitudes, shall be Less than 0.3 $\mu$/s.

16. If Equalizing Pulses are Used in the Vertical Blanking Interval Waveform they shall be 6 in Number Preceding the Vertical Sync Pulse and be at 2H Rate.

FIGURE 7-10. Closed circuit sync waveform.

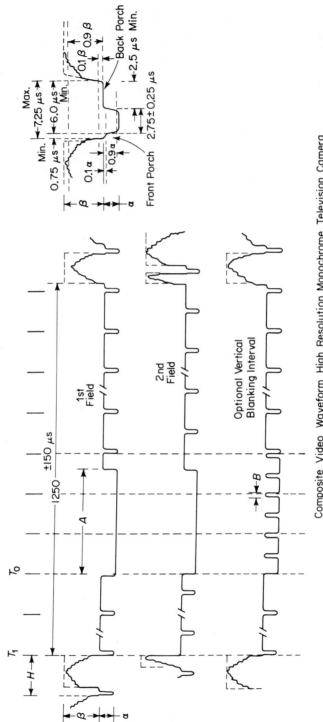

Composite Video Waveform High Resolution Monochrome Television Camera

COURTESY EIA.

FIGURE 7-11. Standard high-resolution sync waveform.

160

per-frame. Here again, the use of equalizing or serration pulses is optional, and this may be attributed to the fact that many modern monitors designed for closed circuit television employ circuitry that is sufficiently sophisticated to operate quite well in the absence of such pulses. It will be observed, however, that many high resolution systems will employ them to assure increased performance.

## SYNC GENERATOR CIRCUITRY

Figure 7-12 is a block diagram representation of a sync generator which would provide the four waveforms necessary to meet EIA Standard RS-170.

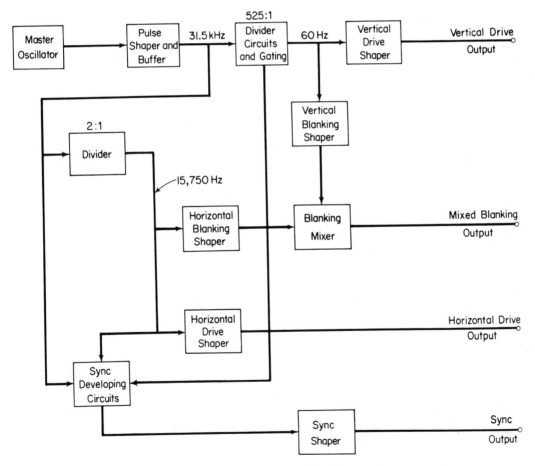

FIGURE 7-12. Block diagram of sync generator.

Synchronizing systems such as the one illustrated are characteristically timed, or driven, by a stable master oscillator. In standard 525-line systems, the frequency of the oscillator is generally 31.5 kilohertz. This is twice the horizontal frequency and is coincident with the frequency of the equalizing and serration pulses that are derived from its output. The buffer stage that is situated at the output of the master oscillator serves the time honored function of isolating the oscillator from the circuitry that it drives, thus assuring that its stability will remain immune to changing load conditions which might be externally imposed. It also serves as an impedance-matching device owing to a high input impedance and an output impedance that is low.

The divider and gating block serves two functions, probably the more important of which is to accurately divide the master oscillator frequency by a factor of 525, thus deriving a 60-hertz output that is phase locked to the original 31.5-kilohertz frequency. This output is used to initiate the Vertical Drive and Vertical Blanking waveforms. The gating portion of the block serves to provide an output that will be used to form the equalizing and serration pulses (by gating other circuits on and off to insert the 31.5-kilohertz signals necessary). Reference to the waveforms shown for RS-170 indicates that such gating action must be coincident with the start of vertical drive and vertical blanking.

The 31.5-kilohertz output from the master oscillator is also fed to a 2:1 divider whose output is 15,750 hertz, the horizontal frequency. This is then used to derive the horizontal blanking, horizontal drive, and horizontal sync signals by its use in the appropriate circuits.

### Master Oscillator Circuits

Figure 7-13 illustrates one type of master oscillator which may be utilized. It is basically an *LC* oscillator whose frequency is primarily determined by *L*1, *C*12, and *C*13. *C*13 is made variable to provide for some adjustment in the frequency. The oscillator may be operated in three different modes. The *free-run* mode allows the circuit to oscillate utilizing only the constants of the *LC* components described. It is in this mode that the oscillator would normally be adjusted to bring the frequency to the optimum operating point. The *crystal* position of the switch inserts a crystal into the circuit and the oscillator will lock to the natural frequency of the crystal. The AFC or *line-lock* mode provides for the application of a dc-

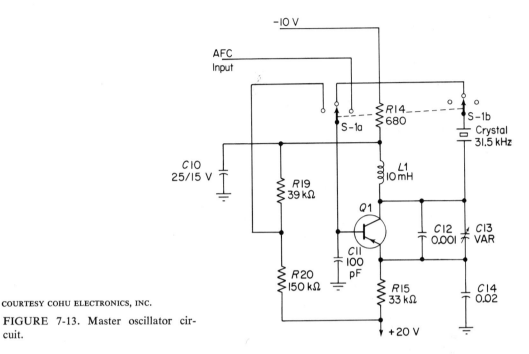

COURTESY COHU ELECTRONICS, INC.

FIGURE 7-13. Master oscillator circuit.

bias voltage to control the oscillator frequency. This voltage is usually derived by comparing the phase of the vertical drive pulse output from the generator with a sample of the power line frequency. Any deviation of the vertical drive phase will create a voltage difference which, when felt at the oscillator, will change the frequency to bring the vertical drive (which has been derived from the master oscillator frequency) back into phase with the power line frequency. In the United States, the power companies take considerable pride in the accuracy of their 60-hertz frequency. They are generally justified in this, as it is probably more accurate than any crystal that would be used in such a circuit.

An example of a voltage controlled oscillator (VCO) is shown in Figure 7-14. The circuit is basically that of a free-running multivibrator whose frequency is also dependent upon a dc input voltage. Because of the multivibrator action, $Q2$ and $Q4$ conduct on an alternate basis. The off-time of $Q2$ is governed by $C2$ and $R3$, and the off-time of $Q4$ is controlled by $C1$ and $R7$. Both of these off-times, however, are also dependent upon the dc voltage applied at the junction of $R3$ and $R7$ through transistor $Q3$. A dc-control voltage is developed in other circuitry and fed to one end of

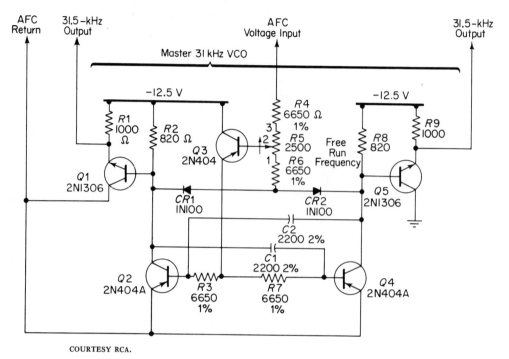

COURTESY RCA.

FIGURE 7-14. A voltage controlled master oscillator.

*R4*. This voltage is attenuated to the proper value for biasing *Q3* by the divider action of *R4*, *R5*, and *R6*. The lower end of *R6* is affixed to a voltage of about −12 volts, which is established by alternate conduction of *CR1* and *CR2*.

Diode *CR1* conducts when *Q2* is off, at which time its collector is at approximately −12 volts. *CR2* conducts when *Q4* is off. These two diodes are included to assure positive starting of the multivibrator at initial turn-on. If both *Q2* and *Q4* were well matched in gain characteristics, it is possible that both could saturate simultaneously at initial turn-on. However, the diodes prevent this, as can be illustrated by assuming that both *Q2* and *Q4* are saturated at the same time. In this condition, the cathodes of both *CR1* and *CR2* would be near ground potential and their anode terminals would also be near ground. If the upper end of *R4* were also near ground (and in this circuit it normally is) the base of *Q3* would, by definition, also be at near ground potential. When this condition exists, neither *Q2* nor *Q4* will be able to turn on, so simultaneous saturation is rendered impossible.

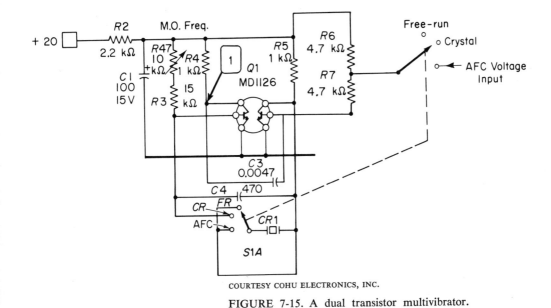

FIGURE 7-15. A dual transistor multivibrator.

Figure 7-15 illustrates a similar type oscillator which may be used. It is a multivibrator that utilizes a dual transistor as the active element. The multivibrator operates essentially the same as in the previous example, but has a *crystal* position as well.

The circuit that generates the bias voltage for the *line-lock* operation of the oscillator in Figure 7-15 is shown in Figure 7-16. Here, $Q1$ acts as a pulse shaper and clipper to provide an output which may be compared to the vertical drive pulse. It utilizes only the negative-going portion of the input sinewave and provides a squarewave output. This output is applied through $C5$ and $CR16$ to one side of the comparator $Q3/Q9$. Vertical drive input is applied to the base of $Q3$ through $CR1$. The two transistors form a bistable multivibrator. Consequently, in order for it to change state it must be triggered.

Assume that $Q3$ is conducting. Since $-10$ volts are applied to the collector-load resistor and $+20$ volts are applied to the emitter through $R62$, the voltage at the collector of $Q3$ will be the result of the dividing action of $R4$ and $R62$. This positive voltage is also felt at the base of $Q9$ through its biasing resistor $R6$, and this will cause it to remain cut off. When a negative-going pulse is applied to $Q9$ through $CR16$, it turns the transistor on and the subsequent positive voltage at the collector turns $Q3$ off by applying a positive potential to its base through $R9$. Thus, the wave-

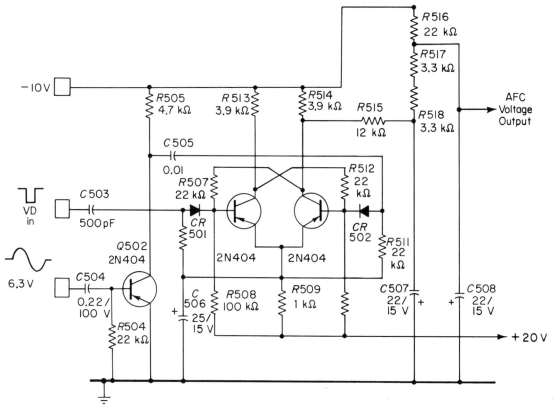

FIGURE 7-16. Circuit generates an AFC voltage by comparing sync generator vertical drive output with the 60-Hz power line frequency.

shape at the output is dependent upon the phase relationship of the pulses at the two inputs. If the output pulse train is applied through a filter, a dc average voltage will result that is dependent upon the relative proportions of the squarewave output.

If the dc voltage obtained from the above filter is applied to the oscillator in Figure 7-15 it will result in the AFC action necessary to lock the generator to the power line frequency. With the vertical-drive output from the generator phase-locked to the line frequency, the output from the comparator will approximate a symmetrical squarewave. Should the phase relationship of the two inputs vary for any reason, the circuit action causes the pulse width to vary and this, in turn, will vary the dc output from the filter. The change in dc will alter the frequency of the oscillator in such a manner that the vertical-drive output signal will shift back into phase with the line frequency.

### Countdown Circuitry

To derive the vertical signals, it is a common practice to count down from the master oscillator frequency to the field rate. To achieve this, circuits of various configuration may be used.

Figure 7-17 illustrates a four-stage counter that will divide the 31.5-kilohertz oscillator frequency down to 60 hertz. Unijunction relaxation oscillators are used to effect the proper frequency division and are arranged so that the four stages divide their input signals by seven, five, five, and three respectively. In operation, each of the stages free-runs at a frequency

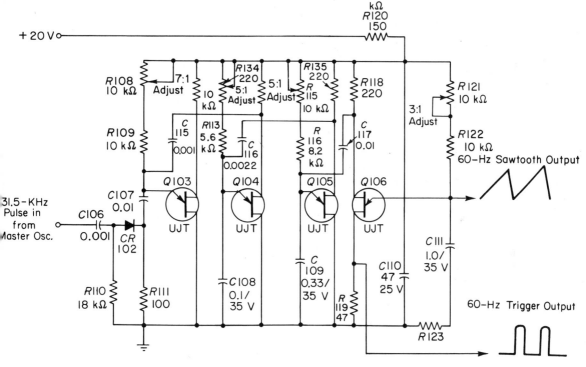

COURTESY COHU ELECTRONICS, INC.

FIGURE 7-17. Countdown circuit uses four unijunction oscillators.

determined by the resistance and capacitance in the emitter circuit. With a series of 31.5-kilohertz pulses applied at the divider input, the potentiometer $R108$ is adjusted until the natural period of oscillation of $Q103$ allows it to be triggered into conduction by every seventh input pulse. The sawtooth at the emitter terminal is therefore one-seventh the frequency of the input, or 4300 hertz. Differentiating the negative-going excursion of this sawtooth and applying it to the following stage, $Q104$, allows this frequency to be divided by a factor of five. Each succeeding stage performs in a similar manner, yielding a total division of 525. (For detailed operation of the unijunction oscillator, see Chapter Five.)

Flip-flop multivibrators are often used as countdown frequency dividers in sync generators. They are popular because of their inherent stability and reliability. Where the counters in the previous example may divide incorrectly if misadjusted or subject to wide extremes of temperature, the *bistable* multivibrator is triggered on and off by succeeding pulses and thus always divides by two.

The integrated-circuit counter shown in Figure 7-18 consists of ten bistable multivibrator circuits, each comprising direct-coupled transistor-gated inputs. In each of the two stable conditions, one of the direct-coupled stages is cut off and the other is conducting. Application of a trigger pulse reverses the conditions. Gates route the trigger only to the stage which is conducting.

The first two counters, $U2$ and $U3$, are connected to form a divide-by-three shift register. The output from this arrangement drives the remainder of the circuits which are connected as normal multivibrators.

The outputs of each binary counter are alternating positive and negative step functions, effectively forming a squarewave output which is one-half of the input frequency. Only the negative step functions will act as triggers on the following direct-coupled binary. Consequently, a binary that has its input directly coupled to the output of the preceding stage completes a cycle, and the result is a division by four over the two stages. Similarly, further cascading of binary counters results in overall division by integral powers of two. The direct coupling of the eight consecutive stages that follow $U2$ and $U3$ would result in an overall division by 256 ($2^8$). Since the total division must be 525, and there has already been a division by three, the remaining division should be 175. To accomplish this, certain counters are reset during specific states. By application of $U11$ output to the reset input of $U4$, $U8$, and $U10$, counter operation is changed so that

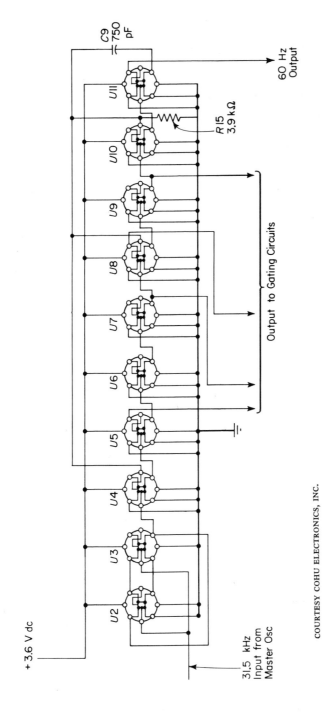

COURTESY COHU ELECTRONICS, INC.

FIGURE 7-18. Integrated circuit multivibrators divide by 525.

175 triggering pulses from the divide-by-three circuit (*U2, U3*) are required at *U4* input in order to produce a second output pulse at *U11*. Thus, with a 31.5-kilohertz input applied to the divider circuitry, a 60-hertz output will be realized. This output can be used to trigger monostable multivibrators which will then provide output pulses at the 60-hertz rate of the proper width for vertical drive and vertical blanking.

Figure 7-19 illustrates a multivibrator that is used to divide the 31.5-kilohertz signal by two to obtain the 15,750-hertz horizontal frequency. The two transistors are connected in an asymmetrical arrangement. The output interval is determined by *R414* and *C403*. Pulse length is adjusted by *R414*, which sets the width to be that of the horizontal blanking pulse needed. An additional output is provided to trigger the horizontal drive and horizontal sync circuits. The time constant set by *R413* and *C403* permits the multivibrator to accept trigger pulses at one-half the master oscillator frequency, thus providing an output that is one-half the frequency of the input.

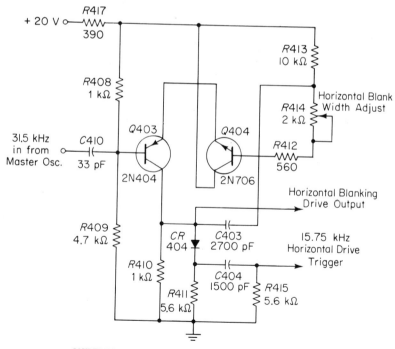

FIGURE 7-19. Multivibrator circuit divides input frequency by two.

## Pulse Shapers and Delay Circuits

To achieve the variety of pulses necessary for an EIA standard sync waveform it is necessary to employ various types of pulse shapers to achieve the necessary pulse widths and pulse delay circuits that are used to establish the rather complex timing relationships.

One of the most popular circuits for achieving the above is known variously as a *boxcar* circuit, *pulse narrower,* or *pulse shaper.* It is basically a simple circuit and is widely used because of its reliability and versatility. Figure 7-20 illustrates a basic boxcar circuit and its associated waveforms. The boxcar circuits derive their name from the *following after* nature of their operation. The leading edge of the output pulse is derived from the trailing edge of the input pulse, and the output pulse width is determined by a simple $RC$ time constant.

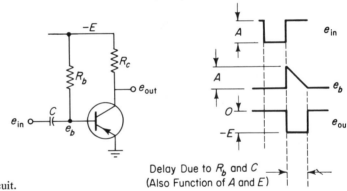

FIGURE 7-20. A "boxcar" circuit.

The circuit in the diagram is normally in a saturated state because of the bias obtained from the power source through $R_b$. Because of the saturated condition, the collector of the transistor is at approximately ground potential. When the leading edge of the negative input pulse is applied to capacitor $C$, the differentiated output serves only to drive the stage further into saturation, and no change in the output occurs. The value of $C$ is such that it can be fully charged during the width of the input pulse. When the trailing edge of the input pulse arrives, the positive-going differentiated pulse that appears at the base drives the transistor out of saturation and the collector voltage shifts abruptly to the supply potential, giving a negative output pulse that has a leading edge precisely coincident with the trailing edge of the input pulse. At this point, the capacitor $C$ is disconnected from ground through the suddenly reverse-biased base-emitter junction of the transistor and is free to discharge through $R_b$. This is indicated in the $e_b$ waveform where the voltage decays toward $-E$ from its initial peak. As $e_b$ reaches ground potential, the transistor is switched on, to return to its original state. Thus, the output pulse width is determined by the values of $R$ and $C$, and the input pulse amplitude. However, the output pulse width is completely independent from the input pulse width. The output pulse width can be made variable by adjusting the value of $R$, $C$, or input amplitude.

If a positive pulse were applied as input to the above circuit, there would be no delay, and the output leading edge would be coincident with the leading edge of the input. The output pulse width would still be independent of the input, however.

Another type of pulse-forming delay device would be the monostable multivibrator with a differentiating network at its input, to allow it to trigger on the trailing edge of the input pulse also. And, of course, adjusting the $RC$ values in the cross-coupling networks will allow an excellent means of varying the symmetry, or pulse width, of the output waveform.

Figure 7-21 shows a circuit that is used to form the horizontal sync, equalizing, and serration pulses, simply by changing the input-pulse level and changing the time constant of the $RC$ network in the base circuit of the pulse shaper. In operation, a series of pulses that are rather low in amplitude and occurring at the horizontal frequency are applied to the input capacitor $C204$. Assuming that $Q204$ is in the off condition, the output pulse at the collector of $Q205$ will be the result of the input pulse and the values of $C204$, $R212$, and $R213$. The output pulse in this case will form the horizontal sync pulse.

If the above circuit is suddenly provided an input of pulses that are considerably higher in amplitude, and that occur at twice the horizontal frequency, it could be expected that the higher amplitude input would result in an output whose pulses were greater in width. This would certainly be the case were it not for the fact that $Q204$ turns on at this same instant, lowering the effective resistance of the $RC$ network by paralleling $R211$ with $R212$ and $R213$. This results in an output of narrow pulses which become the equalizing pulses previously discussed. When the sync gate waveform at the base of $Q204$ turns the switch off, the $RC$ network returns to its previous value and, because the high amplitude pulses are still present, the output pulses are quite wide, yielding the serrated vertical

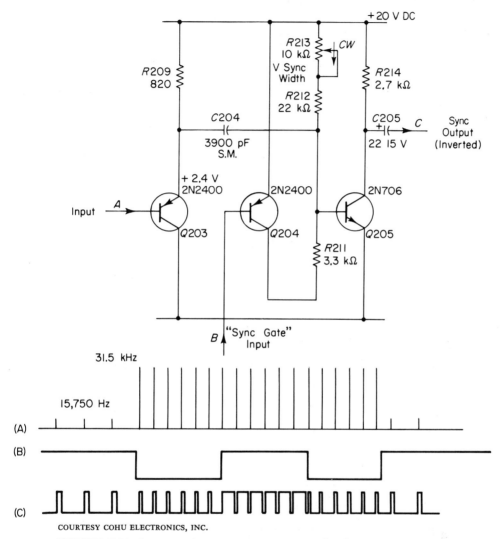

COURTESY COHU ELECTRONICS, INC.

FIGURE 7-21. Sync waveform is generated by altering input amplitude and circuit time constants.

sync interval. Reactivating the *Q*204 transistor switch after this interval results in a series of six more equalizing pulses.

When the last of the 18 high-amplitude pulses has been processed, the lower-amplitude pulses of horizontal frequency are again present at the input and *Q*204 turns off, resulting in a return to horizontal sync output pulses.

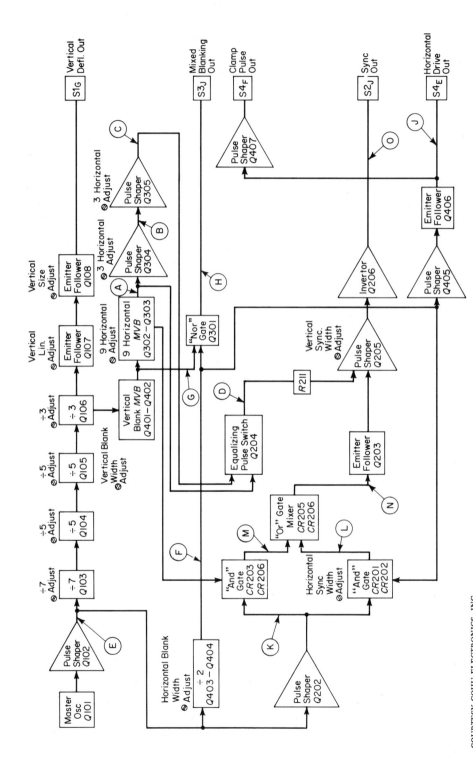

FIGURE 7-22. 525-line sync generator block diagram.

174

From the above, it can be seen that, with proper manipulation of amplitude and resistance values, the simple pulse shaper can be a very versatile circuit indeed.

The sync gate waveform that was used to energize transistor switch $Q204$ may be generated by sampling the proper points of a binary count-down circuit, or may be obtained by use of multivibrators or other pulse-shaping circuits.

## A 525-Line Sync Generator

Figure 7-22 illustrates the complete block diagram of a sync generator used in a closed circuit camera to generate an EIA-type sync waveform for insertion onto the video waveform. Figure 7-23 shows the waveforms generated within the circuitry and indicates the points in the block diagram where they occur. The circuitry used within the diagram consists of many of the circuits previously described.

The master oscillator output at $(E)$ occurs at a frequency of 31.5 kilohertz. It is divided down to the vertical rate by the four indicated counters, and the 60-hertz output, in addition to driving the vertical deflec-

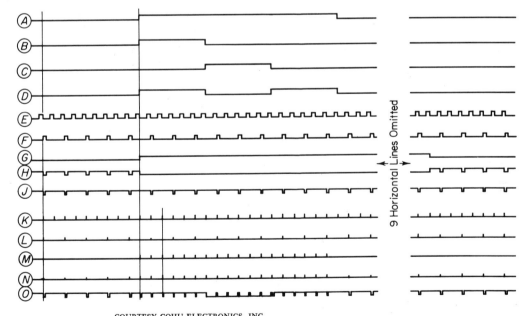

COURTESY COHU ELECTRONICS, INC.

FIGURE 7-23. Waveforms keyed to block diagram of Figure 7-22.

tion stages, also triggers the vertical blanking multivibrator, whose output waveform can be seen at $(G)$.

The 31.5 kilohertz is also applied to the horizontal blanking multivibrator, where it is divided by two and provides the horizontal blanking pulse output $(F)$, and to pulse narrower $Q202$ where it is used to drive two AND gates $(K)$. AND gate $(CR203/CR204)$ receives another input from the $9H$ multivibrator which, in conjunction with the 31.5-kilohertz input, provides a burst of 18 pulses at the 31.5-kilohertz rate once during each vertical interval $(M)$. These pulses are added to those at $(L)$ to provide the waveform that will drive the pulse narrower $Q205$ $(N)$.

The pulse narrower $(Q205)$ and equalizing pulse switch $(Q204)$ operate as shown in Figure 7-21 and provide an output which, after being inverted, is shown at $(O)$.

Horizontal drive is developed by using the leading edge of the horizontal blanking pulse $(F)$ to trigger pulse narrower $Q402$ to obtain $(J)$.

With the advent of the integrated circuit, the physical size and circuit complexity of synchronizing generators has been reduced considerably. Because of the increasing capability of such circuits, it seems very likely that a relative few of these on a miniature circuit board will soon be all that is necessary to comprise the most sophisticated of synchronizing circuitry. It will then be possible to provide a large majority of the camera systems with accurate and stable synchronizing systems, with a relatively small increase in cost.

# MONOCHROME TELEVISION
# MONITORS

Due to the widespread familiarity with the home television receiver, the television monitor's operating principles are perhaps more generally understood than any other single unit in a television system. The monitor differs from the television receiver only in that it does not contain the RF and IF sections which are necessary to receive television signals that have been modulated onto carrier frequencies for broadcasting purposes. It may or may not employ audio circuitry. In these respects the monitor may be thought of as a somewhat simpler device. However, the performance specifications and video presentations of many monitors far surpass that which is obtainable from home television receivers. (Receiver circuits are covered in Chapter 14.)

The television monitor is normally classified as to its bandwidth capability, scanning rate, and picture tube size. The particular type that might be used in an application is generally dependent upon the camera system in use. For example, when utilizing a camera that has a 10-megahertz bandwidth, it is desirable to employ a monitor whose bandwidth capabilities equal or exceed this figure. The scanning rate is chosen to match that of the camera deflection system, and the picture tube size is selected for convenience of viewing.

## GENERAL DESCRIPTION

Figure 8-1 is a simplified block diagram of a typical television monitor.

The input to the monitor is a composite signal consisting of video and sync components. The video signal provides the kinescope with infor-

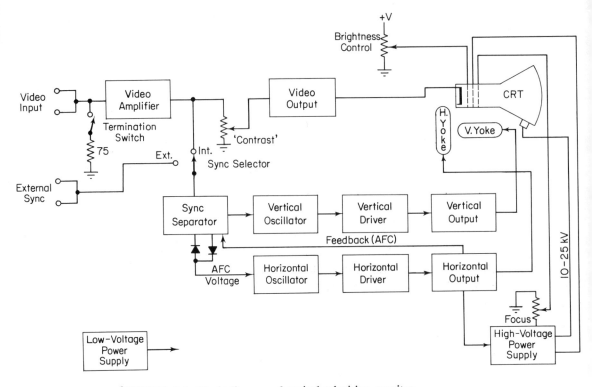

FIGURE 8-1. Block diagram of typical television monitor.

mation to compose the picture, and the sync signal synchronizes the kine-scope electron-beam scanning with the camera scanning.

Many monitors will accept a video input which may be either com-posite or noncomposite. If sync is not present on the video waveform, it must be added at the external sync input (in this case), and the sync selector switch placed in the *external* position. The video input can usu-ally be terminated in the proper resistance value by use of a termination switch provided at the input or, when it is desired to "loop-through" the video to some other terminated destination, the switch may be left open.

When the incoming video signal is composite, both sync and video information are processed and amplified by the first video amplifier. At the output of the amplifier a gain adjustment, normally called the *contrast* control, is incorporated to vary the signal amplitude to the video output stages. Increasing the video amplitude will result in a change in contrast ratio on the kinescope.

The signal output from the first video amplifier is also routed to a sync separator circuit (when the sync selector switch is in the *internal* position). Here the sync information is extracted from the video signal and the horizontal and vertical components of the sync signal are developed. The vertical sync information is applied to the vertical oscillator as a trigger to synchronize it to the incoming signal. Both the vertical and horizontal oscillators are free-running oscillators, and the incoming signals serve simply to lock them to the proper frequency.

The horizontal component of the sync signal is converted into an AFC voltage by diodes $D1$ and $D2$ and applied to the horizontal oscillator. Feedback is provided from the horizontal output circuitry and referenced to the horizontal sync, causing a shift in the AFC voltages when any frequency deviation occurs in the horizontal oscillator. The change in voltage is applied to the oscillator and serves to pull it back onto frequency.

The kinescope beam is magnetically deflected by use of magnetic fields produced by horizontal and vertical coils in the deflection yoke. The electron beam deflection is proportional to the current in these coils. Output from the vertical oscillator is fed into the vertical driver block which provides the necessary power to drive the vertical output stage. The horizontal driver accomplishes essentially the same thing for the horizontal deflection circuitry, but where the vertical output had only the yoke as a load, the horizontal output is also generally utilized as drive for the high-voltage power supply.

The high-voltage power supply transforms the horizontal output pulses into pulse excursions of extremely high amplitude. These are rectified into a dc voltage of several thousand volts to provide the ultor voltage for the kinescope. Generation of the high voltage from the horizontal frequency was originally utilized because of the difficulties in shielding and filtering the high potentials involved. Thus, any undesirable effects arising in these problem areas are much less noticeable since they are synchronous with the horizontal deflection.

The picture tube accelerating grid and focus element voltages are also generally derived within the high voltage section.

## THE KINESCOPE AND YOKE ASSEMBLY

The kinescope, or picture tube, is a cathode ray tube that essentially converts electrical signals into light patterns. Electron emission from a heated cathode is formed into a narrow beam of fast-moving electrons and impelled toward the phosphor coating that covers the inner surface of the faceplate (Figure 8-2). As the electrons strike the phosphor there is a release of energy which excites the phosphor and causes visible light to be emitted. The color of the light is determined by the type of phosphor that is used, and the brilliance, or brightness, is a function of the amount of beam current allowed to strike the phosphor coating and the speed at which the electrons are traveling at the moment of impact.

The beam within the kinescope is generally modulated by applying the video signal at its cathode. Such action regulates the amount of beam current, which, in turn, varies the spot brightness on the phosphor coating by an amount and rate determined by the video signal content. It is also possible to control beam current by varying the Grid #1 potential, and it

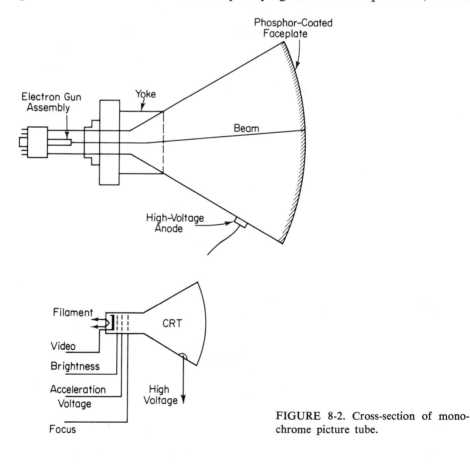

FIGURE 8-2. Cross-section of monochrome picture tube.

is general practice to use a variable bias control in conjunction with $G1$ to give a manual adjustment for brightness.

Some monitors route blanking signals (in addition to those present on the cathode video signal) to $G1$ to assure positive beam cutoff during retrace. $G2$ provides accelerating action for the beam, and $G3$ is made voltage-variable for control of beam focus. The high-voltage anode, or ultor, assures the high beam velocity necessary to ignite the phosphor. Typical voltage for the ultor generally lies between 12 and 25 kilovolts, depending upon tube type and size.

Figure 8-3 illustrates a means of obtaining kinescope blanking by means of negative pulses applied to the $G2$ element. This method takes vertical and horizontal waveforms from their respective deflection circuits and uses them at the screen grid to interrupt the accelerating voltage, thus disabling the beam during retrace. The vertical waveform is obtained from the secondary of a vertical output transformer and coupled through $C242$ to the cathode of diode $X201$. The configuration of the diode within the circuit clamps the output waveform to a fixed potential. The diode conducts up to a specified limit and produces the negative-going waveform indicated. The horizontal waveform comes from a winding on the flyback transformer in the horizontal deflection circuitry. The waveform is applied

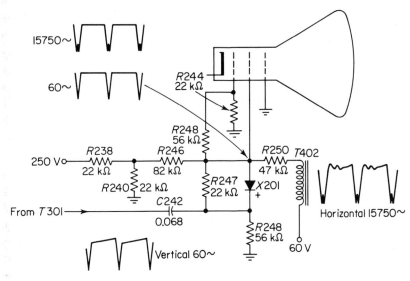

FIGURE 8-3. Horizontal and vertical blanking are applied to the G2 element.

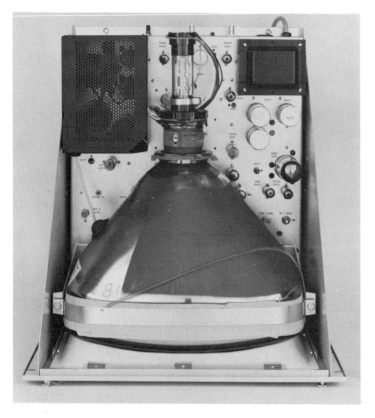

COURTESY COHU ELECTRONICS, INC.

FIGURE 8-4. Top view of closed circuit monitor showing yoke placement on picture tube.

to the anode of diode $X201$, which clips the top portion of the waveform and provides the negative-going pulses that occur during horizontal retrace. Thus, the waveform indicated at $G2$ is a combination of negative pulses occurring during horizontal and vertical retrace intervals, which cuts off the beam to achieve blanking.

Almost all monitor kinescopes employ a yoke around the neck of the tube to realize magnetic deflection (Figure 8-4). The principles of operation of the monitor yoke are essentially the same as that of the camera yoke, although the physical configuration is made quite different in order to achieve proper deflection through the wide angles encountered in the display device. Sawtooth current waveforms are used for drive, and power consumption is relatively high due to the wide scanning excursions necessary. Focusing of the beam is not achieved magnetically but is accomplished by use of the electrostatic focusing elements within the tube envelope.

## VERTICAL DEFLECTION CIRCUITRY

Figure 8-5 shows a schematic representation of one method of achieving the vertical deflection waveform necessary to drive the vertical deflection windings of the yoke. $Q1$ is a unijunction oscillator whose free-running frequency is controlled by the $R$-$C$ time constant in its emitter circuit. When the voltage at the emitter of the unijunction reaches the turn-on point, the charge on the series capacitors $C4$ and $C5$ will be drained to ground through the emitter-base 1 junction. When the voltage at the emitter reaches approximately two volts, $Q1$ turns off and the exponential rise in voltage at the emitter begins again. The result is a sawtooth waveform at the emitter. $P1$ controls the charge time on the capacitors and is

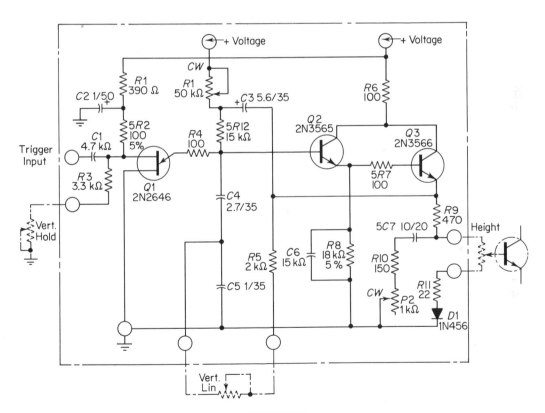

COURTESY CONRAC CORPORATION.

FIGURE 8-5. Vertical driver circuits for monitor deflection.

the *frequency* adjustment. This control is normally set to a point where the free-running frequency will be slightly lower than the vertical deflection frequency, thus allowing positive triggering action by the incoming sync. The frequency adjustment is normally an internal adjustment on most monitors and not available on the front panel.

$Q1$ is triggered at its base-2 terminal by negative-going pulses which have been derived from the vertical sync pulses. The amplitude of these pulses is determined by the *vertical hold* potentiometer which is normally located on the monitor front panel. Amplitude is adjusted for positive locking action at the vertical oscillator. The negative pulse thus applied to the base-2 serves to cause conduction in the unijunction somewhat sooner than would have occurred due to the natural time constants in the emitter circuit.

The sawtooth waveform developed at the emitter of $Q1$ is applied to the base of $Q2$, an emitter follower which, in turn, drives emitter follower $Q3$. Linearity correction is achieved by positive feedback from the emitter of $Q3$ to the base of $Q2$ through the *vertical* potentiometer. The collectors of $Q2$ and $Q3$ are tied together for additional linearity correction.

The linearized sawtooth at the emitter of $Q3$ is applied through the *height* control to the base of the vertical output transistor (Figure 8-6). Adjustment of the height control determines final output amplitude of the

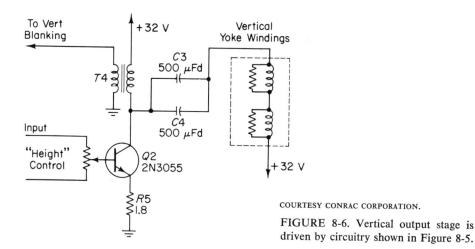

COURTESY CONRAC CORPORATION.

FIGURE 8-6. Vertical output stage is driven by circuitry shown in Figure 8-5.

vertical deflection waveform and sets the effective height of the raster as displayed on the kinescope.

Transformer $T4$ couples the sharp negative-going transition of the sawtooth to additional circuitry to generate vertical blanking information.

Figure 8-7 illustrates a similar method of achieving vertical deflection. In this case the input trigger signal is positive-going. $Q1$ inverts the vertical sync information and applies the resulting negative pulse as triggering for oscillator $Q2$.

It should be noted that the output from $Q2$ is taken from the $B1$ lead. The pulse present at this point is a positive, narrow excursion which is used to turn on transistor switch $Q3$ when the unijunction, $Q2$, is on. During this short period of time the capacitors at the base of $Q4$ are discharged to approximately ground potential. Thus, when $Q3$ ceases conduction, the capacitors will begin to charge exponentially until $Q3$ again conducts. This action forms a sawtooth at the base of $Q4$. The sawtooth that was formed at the emitter of the unijunction oscillator, $Q2$, is not used as the deflection waveform to avoid loading effects. The oscillator will thus maintain frequency stability throughout load variations caused by changing height or linearity controls.

The amplified output of $Q4$ is applied to emitter follower $Q5$ through the *vertical height* control, $R14$. $Q5$ provides the necessary power gain to drive the vertical output transistor, $Q6$. Linearity in this circuit is provided by sampling the output directly at the collector of the output transistor and providing feedback through the *vertical-linearity* potentiometer to the base of $Q4$. The bias adjustment $R20$ serves to correct for nonlinearity at the top of the raster by controlling the bias current of the output transistor.

In Figure 8-8, a vertical oscillator is shown which is in the configuration of a blocking oscillator. Again, the oscillator is designed to have a free-running frequency that is slightly under 60 hertz, and is synchronized by pulses from the sync separator circuitry.

In operation, the blocking oscillator remains cut off, or "blocked," for most of the cycle by a negative voltage applied at the base. The transistor starts conducting when $B$ plus is applied, routing a positive voltage to the base and collector. Current flows through the primary of $T300$ up to the time of saturation, charging $C310$ to a negative potential. When the saturation point is reached, there is no further change in the amount

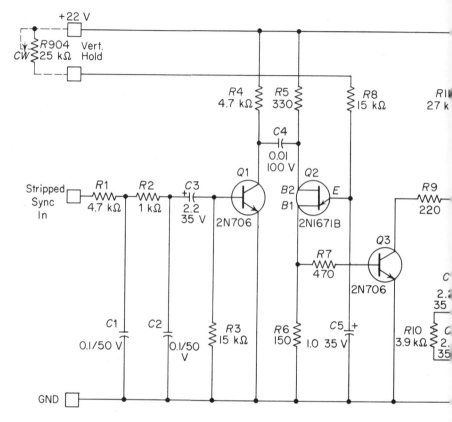

FIGURE 8-7. Triggered unijunction oscillator drives sawtooth generator for vertical deflection.

of current flowing through $T300$. The magnetic field around the primary therefore ceases to change, and the current being induced in the secondary ceases. The negative voltage on the capacitor, $C310$, causes the transistor to turn off, and the collapsing magnetic field that results around the primary assures a very positive turn-off. Capacitor $C310$ discharges slowly through $R325$, keeping the base at a negative potential. As soon as $C310$ has discharged below the cutoff point, the transistor once again conducts and another cycle is started. Synchronization of the oscillator is achieved by applying a positive pulse to the input through $R322$. The output waveform is a negative pulse felt at the collector during the time that the transistor is in the on condition.

Diode $X306$ is important because it conducts immediately after the collector current has reached the saturation point. At this time the field

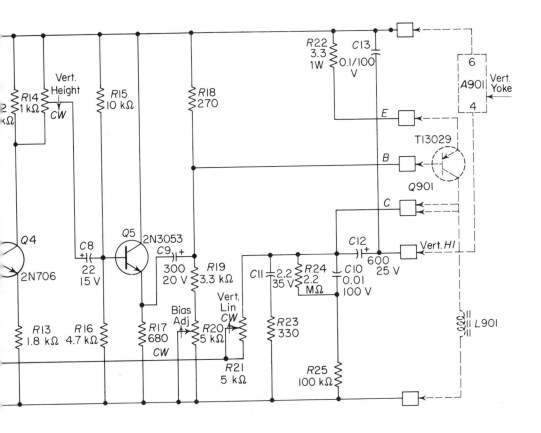

around the primary collapses and reverses direction, placing a more positive potential at the collector of the oscillator than was possible with only the applied voltage. This is an undesirable condition because the transistor could conduct again. However, the diode removes this possibility by effectively damping the positive transition.

A vertical output stage designed to be driven from the foregoing blocking oscillator appears in Figure 8-9. The negative-going pulse from the oscillator is used to drive the capacitor to a reduced potential. It then begins a slow recharging through resistors $R318$, $R334$, and $R332$, resulting in a sawtooth waveform at the base of $Q304$. Two shaping adjustments are provided, $R335$ for overall linearity correction and $R331$, which adjusts the linearity at the top of the picture. $Q304$ acts as an emitter follower to achieve the current gain necessary to drive the output stage, $Q305$. The

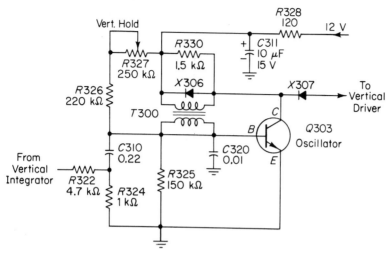

FIGURE 8-8. A blocking oscillator used for generating vertical deflection.

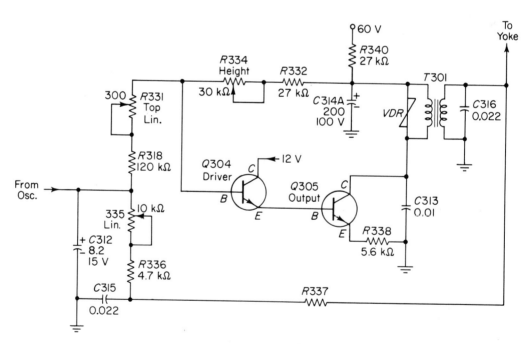

FIGURE 8-9. A vertical output stage for use with the oscillator of Figure 8-8.

output is applied to the vertical deflection yoke through the vertical transformer, $T301$. The voltage-dependent resistor in the primary of the transformer serves to protect the circuit elements from high-amplitude pulses which might be caused during field collapse when the retrace interval occurs.

## HORIZONTAL DEFLECTION AND
## HIGH-VOLTAGE GENERATION

The horizontal deflection circuitry is generally considerably different from that used for vertical deflection, primarily because of the higher frequencies involved. Heavy demands are made upon the horizontal output transistors because of the high current levels necessary and the large amplitude retrace voltages generated by the collapsing magnetic field at the end of each horizontal scan. In addition, most television monitors derive their high voltage potentials by utilizing a portion of the horizontal output signal to drive a high-voltage transformer.

The vertical output stage saw the yoke as primarily a resistive load, but the higher horizontal scanning frequency forces serious consideration of the inductive effects of the horizontal windings of the yoke and other inductive components in the circuit. Output transistors with fast switching speeds are necessary to minimize power dissipation within the semiconductor elements during the actual switching period, and adequate heat sinks must be provided to conduct generated heat away and maintain a stable temperature.

It is also interesting to note that the horizontal oscillator in most modern monitors is not directly triggered by the horizontal sync pulses. Although some early monitors (and television receivers) utilized this method, they were found to be susceptible to false triggering by noise spikes that might find their way through the sync separator circuitry. For this reason, the horizontal oscillator is generally controlled by an AFC system which compares the phase of the horizontal output with that of the incoming sync pulses. If any deviation occurs between the two, a dc voltage is generated and applied to the horizontal oscillator to alter its frequency so that a constant phase relationship is maintained. Such a system is relatively immune to spurious signals and a more stable mode of operation is achieved.

### Horizontal Oscillators and Drivers

Figure 8-10 illustrates an AFC circuit and horizontal oscillator. The active elements in the AFC circuit consist of diodes $X405$ and $X406$, and transistor $Q403$. The diode pair requires two inputs to generate a correction voltage to control oscillator frequency. One input consists of negative-going pulses derived from the sync separator circuits, and is applied to the anodes of the diodes through $C412$. The other input is a sawtooth waveform obtained by sampling the horizontal deflection waveform. The two pulses combine in the diode network to generate a correction voltage. The negative pulses from the sync separator serve to turn the diodes on and off. As the phase between the sawtooth waveform and the sync input varies, the point at which the diodes are switched will vary along the slope of the sawtooth, thus causing a change in the average dc component at the junction of $R421$ and $R422$. This is filtered and applied to the base of $Q403$ as a dc bias.

$Q403$, the AFC amplifier, serves as a simple dc amplifier, converting the small changes in dc potential at its base into larger voltage excursions

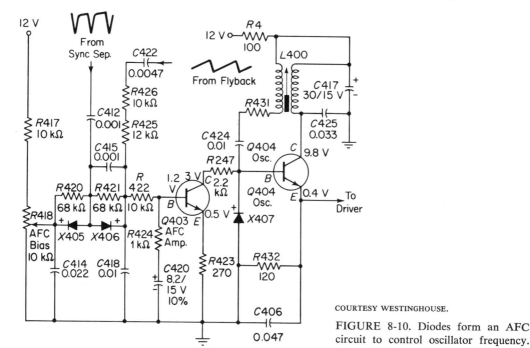

COURTESY WESTINGHOUSE.

FIGURE 8-10. Diodes form an AFC circuit to control oscillator frequency.

at its collector. These larger voltage changes are applied to the oscillator as bias that will alter the natural period of oscillation.

The oscillator is a blocking oscillator, in slightly different configuration than that used in the previous example for the vertical oscillator. Its principles of operation are generally similar, although in this case the natural period of oscillation is determined by the value of $C424$, the presence of diode $X407$, and the value of dc potential felt at its base. Thus, as the AFC amplifier varies the base voltage when a deviation in frequency occurs, the oscillator will be pulled back into the proper phase relationship with the reference horizontal synchronizing waveform.

Figure 8-11 shows another type of oscillator followed by the succeeding driver stages necessary to provide the proper drive signals to the horizontal output stage.

$Q1$ and $Q2$ form a multivibrator whose natural frequency of operation is approximately that of the horizontal frequency of the incoming signal. A manual control of frequency is achieved by applying bias to transistor $Q1$ through the *horizontal frequency* control. This voltage is obtained from a source that is regulated by the action of Zener diode $D2$, thus assuring stable operation. An input for AFC voltage is also provided through resistor $R1$ to the base of $Q1$, allowing the oscillator to maintain the stable phase relationship with the horizontal sync as previously discussed.

Capacitors $C8$ and $C9$ serve as filtering elements for the AFC voltage. Transistors $Q3$, $Q4$, and $Q5$ are current amplifiers to provide sufficient drive to the horizontal output stage. $Q3$ and $Q4$ operate as emitter followers to provide isolation for the oscillator and allow sufficient current gain to control the transistor switch $Q5$. $Q5$ has a transformer primary as its load which couples the output pulse to the horizontal output transistors.

### Horizontal Output Circuitry

Figure 8-12 shows a simplified diagram of a horizontal output stage. The two horizontal output transistors are parallel driven through the action of the transformer secondaries connected to their respective base terminals. The transistors are connected in a series arrangement that allows them to share the voltage transitions that occur during retrace, thereby dividing the power-handling capabilities necessary in the individual transistor

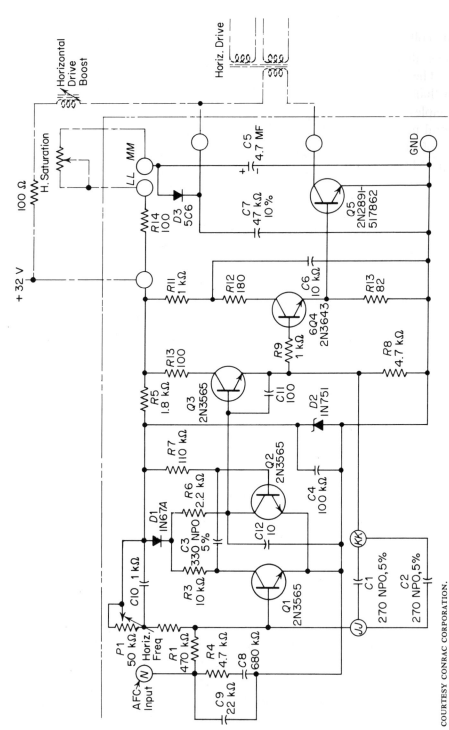

FIGURE. 8-11. Horizontal drive signal is initiated by a voltage controlled multivibrator.

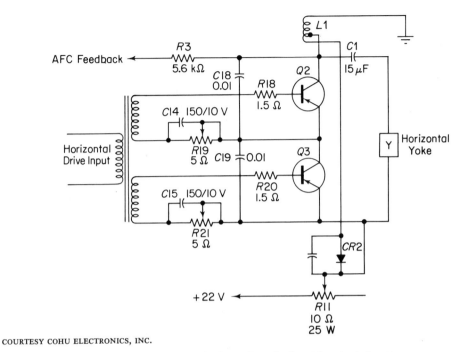

FIGURE 8-12. Simplified diagram of a dual transistor horizontal output stage.

elements. Advances in semiconductor technology are rapidly eliminating the need for series-connected output stages, but their use is presently demanded where relatively large screens and high scan rates are necessary.

The two horizontal output transistors are series-connected switches that open and close simultaneously. The current buildup in the yoke is not instantaneous however, due to its inductive nature. The current increases in a relatively slow manner, forming a sawtooth current waveform. The capacitive and inductive components in the circuit are usually chosen to create a time constant sufficiently long enough to allow utilization of only the initial part of the sawtooth curve. This assures use of the most linear portion of the waveform, and, since the spot position of the kinescope electron beam is almost directly proportional to the current in the deflection yoke, results in a sweep pattern with good linearity characteristics.

During scan, $C1$, the yoke, and the series inductors form a resonant circuit. The first half of the scan is produced by energy stored in the yoke from the retrace action of the previous scan excursion. The current from the yoke charges $C1$ through Diode $D2$. During this period, the current is increasing in the yoke and its series inductors, causing the beam to be deflected through the first half of its scan deflection.

The second half (approximately) of the scan is produced by discharging $C1$ through the yoke with $Q2$ and $Q3$ completing the circuit as a closed switch. The voltage across the yoke and series inductor elements is maintained so the current continues to increase. When the switch transistors open at the end of the scan, the yoke inductance, now paralleled by the series inductor, resonates with $C2$ at approximately 60 to 100 kilohertz (depending upon the component values used). The energy stored in the inductances causes the circuit to ring for half a cycle to produce retrace and reverse the current in the yoke. Diode $CR2$ is back-biased during the first half-cycle of the ringing waveform but comes into conduction when the waveform starts to go positive, damping any further ringing.

The reverse current through the yoke causes the beam to retrace rapidly, as a function of the high ringing frequency. When retrace has been accomplished, and the damper diode begins to conduct, the series inductance is again in the circuit and the resonant frequency of the circuit is reduced to its original lower value to allow normal scan operation.

### High-Voltage Generation

The ultor voltages necessary to properly operate a kinescope tube are generally derived from the horizontal output signal. The familiar "flyback" transformer found in nearly all receivers and monitors transforms the horizontal retrace pulse that appears across its primary winding into the high potentials necessary. A diode in the secondary circuit rectifies the transformer output and a capacitor-resistor network is usually used as filtering. Since the current drain is quite small, large values of capacitance are unnecessary.

Figure 8-13 depicts the circuitry necessary to form the high voltages and illustrates the interaction between the horizontal output and high voltage sections. The heavy lines indicate the path that the retrace current takes as the yoke discharges during the retrace interval. It can be seen that the primary of the transformer, $T2$, is in series with the yoke. During retrace, energy is coupled into the secondaries, which are wound to effect a voltage step-up. Diode $V1$ rectifies the secondary output and applies high voltage to the kinescope anode through $R6$, with filtering being accomplished by capacitor $C7$. Other windings of the transformer yield the lower voltages that are necessary for the accelerating and focusing grids.

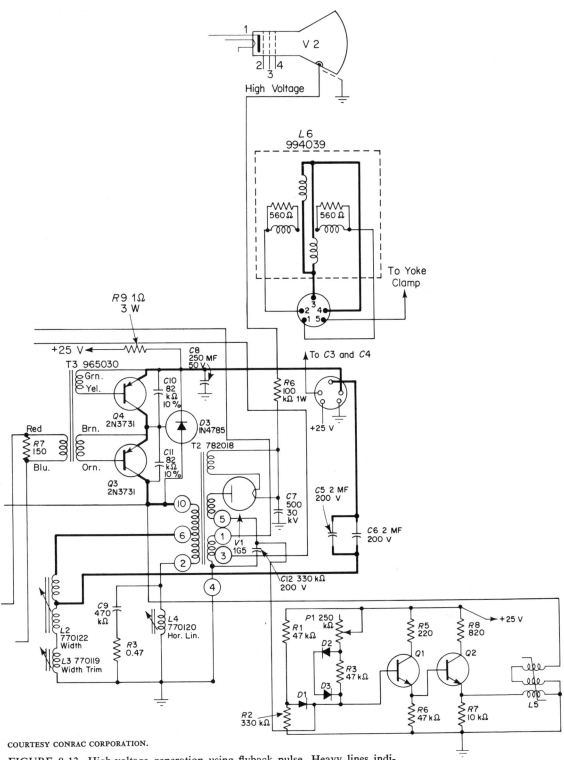

FIGURE 8-13. High-voltage generation using flyback pulse. Heavy lines indicate retrace current.

195

High-voltage regulation is necessary to eliminate variations in picture quality as brightness or contrast controls are changed, or wide variations in overall brightness levels are encountered in televised scenes. One method of accomplishing regulation is to employ a saturable reactor ($L5$) in parallel with the flyback primary winding. In operation, as the kinescope anode current increases (as it would with higher brightness levels), the regulator assembly senses this increased current through the bottom of the tertiary (HV secondary) winding. $Q3$ and $Q4$ amplify the current to drive the control winding of the saturable reactor. As the control winding current increases, the saturable winding inductance is reduced, thereby allowing the winding to store more energy during trace time. This stored energy, delivered to the flyback transformer during retrace time, produces additional voltage at the plate of the high-voltage rectifier, $V1$, thus maintaining the kinescope anode voltage at a constant level.

### Linearity and Width Control

Figure 8-13 also illustrates a novel means of obtaining control over the horizontal width of the monitor raster. The horizontal width control, $L2$, employs a series and a parallel coil coupled with a movable core. Moving the core changes the inductance of the device, and since one-half of the coil is in series with the yoke, the effective inductance of the entire horizontal circuit is changed. This changes the current waveform during scan and causes a corresponding change in horizontal size. As the width is varied by moving the core, the impedance of one-half of the $L2$ winding is increased while the impedance of the other one-half is decreased. This allows a variation in width while presenting a constant load to the flyback and, therefore, maintaining a constant high voltage.

Horizontal linearity is achieved with a resonant tank circuit in series with the flyback primary ($L4$ and $C9$). This circuit adds a sawtooth and a parabolic component to the sawtooth current in the yoke. $L4$ determines linearity by controlling the amount and shape of the correcting voltage waveform.

## SYNC SEPARATION

It is the purpose of sync separator circuits in a monitor or receiver to derive from the video signal those components that have been included for synchronizing the horizontal and vertical deflection systems with those of the camera. Most separator circuits remove the video portion of the composite waveform by selective biasing of the input amplifiers. The sync waveform that is retained is then separated into horizontal and vertical drive pulses. The horizontal pulse is generally processed into an AFC voltage to control the horizontal oscillator (as previously demonstrated), while the vertical pulses are applied to the vertical oscillator as direct triggering. A detailed explanation of the principles of sync separation is covered in Chapter Seven.

Figure 8-14 illustrates an example of sync separator circuitry. The composite signal from the video amplifier board is coupled by $C706$ to emitter follower $Q704$. The emitter follower serves to isolate the sync separator input. The output of $Q704$ drives a clipper circuit, $Q705$, which is biased so that only the sync component of the composite video signal causes it to conduct. The video signal is therefore eliminated, and the amplified sync component is fed into an integrating network for derivation of the vertical drive and to the differentiating network consisting of $C707$ and $R717$.

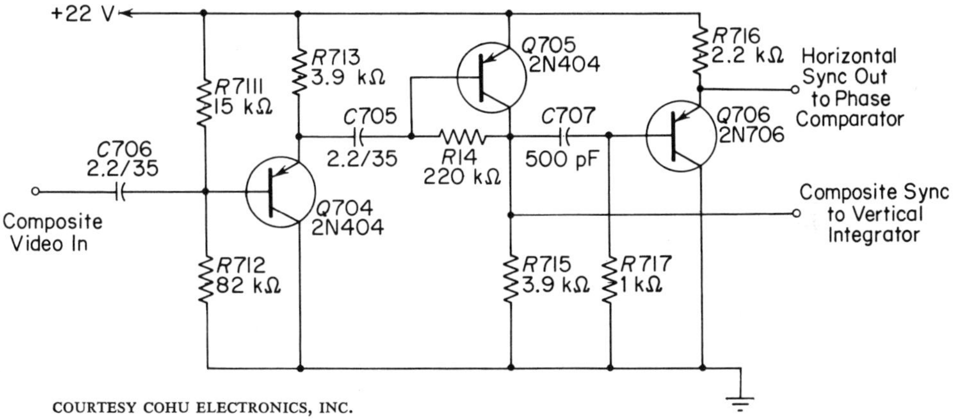

FIGURE 8-14. Sync separator circuit.

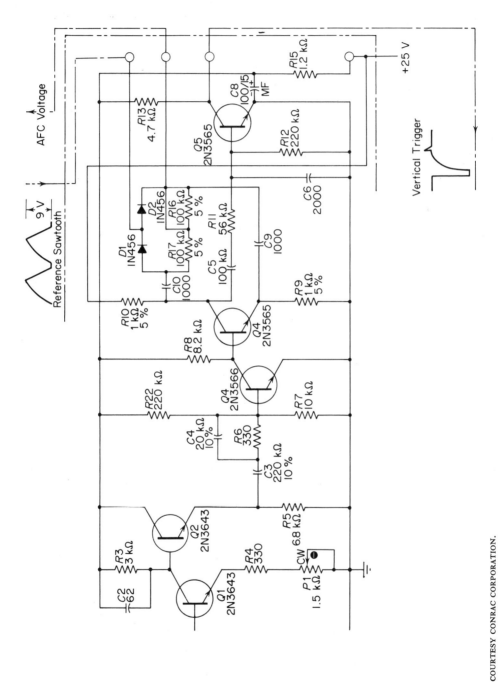

FIGURE 8-15. Sync separator and horizontal AFC voltage generator.

At this point the sync signal consists of periodic groups of equalizing pulses, serrated vertical pulses, and horizontal sync pulses, all of which (in this case) are positive-going in polarity. The differentiating network, $C707/R717$, reduces all pulses to the same width, and provides an output consisting of a positive pulse and a negative pulse for each input pulse. The differentiator output is applied to $Q706$ which is turned on by the positive voltage excursions. The output is a series of negative-going pulses that are fed into the phase comparator circuit. Here the pulses are compared in phase with the horizontal drive being applied to the yoke. The comparator output is a dc voltage with a polarity and level that are determined by the sign and magnitude, respectively, of the phase difference between the synchronizing pulses and the horizontal output signal.

Figure 8-15 illustrates a slightly different means of achieving sync separation and again illustrates the operation of the comparator. Here $Q1$ and $Q2$ amplify the composite video (or sync signal where external sync is used) to drive the sync clipper, $Q3$. The biasing of $Q1$ and $Q2$ is such that the video signal is largely eliminated. The clipped sync at the output of $Q3$ is applied to the phase splitter, $Q4$, which, in turn, drives the phase comparator consisting of $D2$ and $D3$. The positive and negative pulses applied to the diodes serve to turn both on simultaneously to sample the sawtooth that is present at their junction. The point where this occurs along the sawtooth ramp determines the polarity and magnitude of the dc voltage output.

The output from the collector of the phase splitter is also coupled into $Q5$ which has an integrating network in its base circuit that develops the vertical sync pulses to trigger the stage. The output of $Q5$ is a negative-going pulse which is used to trigger a vertical oscillator.

## VIDEO PROCESSING AND AMPLIFICATION

Video amplifiers in monitors (and receivers) function to amplify the incoming video signals to an amplitude sufficient to drive the kinescope cathode for proper beam modulation. In general, the output signal level demanded may vary anywhere between 10 to 100 volts, depending upon input signal amplitude, setting of the contrast control, type of kinescope, and the like. In addition to simple amplification, the video amplifier must have good high-frequency characteristics to faithfully reproduce the fine detail elements that may be contained in the video signal. Since detailed resolution depends upon the bandwidth of the amplifying system, it must be broad enough to pass those frequencies that represent the smallest picture elements desired.

Commercial television receivers are designed with video bandwidth characteristics of something less than 5 megahertz because limitations of the transmitting bandwidths of commercial television stations limit video signals at the receiver to about 4.5 megahertz. Any additional bandwidth in a commercial receiver would serve only to amplify noise in the area above the 4.5-megahertz frequency. In many closed circuit television systems, however, video bandwidths that extend to 20 or 30 megahertz are not at all uncommon. Figure 8-16 illustrates a bandwidth plot which shows some of the common bandwidth capabilities normally encountered.

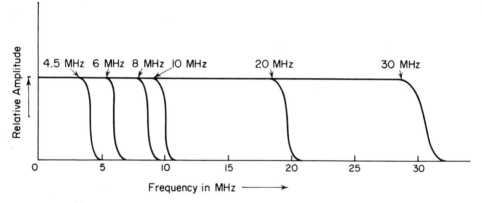

FIGURE 8-16. Some typical bandwidth curves for closed circuit monitors.

To meet the requirements of video amplification it is necessary to utilize an amplifier whose amplitude and phase response are essentially flat from dc to the upper bandwidth limit. Since capacitive coupling is generally employed in amplifier circuits it is, of course, not possible to actually achieve response to the dc component of the signal. Therefore, the lowest frequency that must be considered is that of the horizontal scanning frequency. The dc component of the signal can be restored with additional circuitry at the amplifier output.

Figure 8-17 shows an amplifier with a 10-megahertz bandwidth. Although the input is capacitively coupled from the contrast control, the circuitry consisting of $Q2$, $Q3$, and $Q4$ functions as a dc amplifier. The

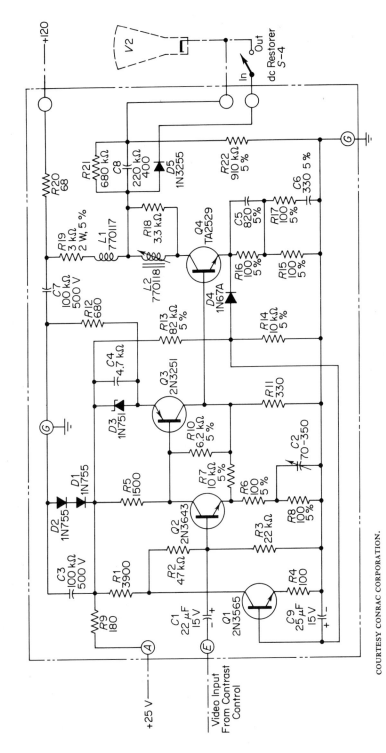

COURTESY CONRAC CORPORATION.

FIGURE 8-17. A monitor video amplifier with 10-MHz bandwidth.

201

operating point of the amplifier is controlled by a clamping circuit which consists of active elements $D4$ and $Q1$. The clamping circuit uses the sync tips as a reference. Since the sync waveform on a composite video signal does not vary in amplitude, this gives the amplifier a stable reference point and allows it to restore the dc component of the signal which was lost due to capacitive coupling in the earlier stages. With no input signal, the three amplifier transistors operate at a minimum current. When a composite video signal is applied, the negative-going sync pulses at the emitter of $Q4$ cause $D4$ to conduct, which results in increased collector voltage of $Q1$. This, in turn, raises the current operating point of the amplifier to a new level, yet maintains the sync tips at the same dc level. The time constant involved in the system is long enough to allow the lowest usable frequencies to be displayed. The negative feedback provided by the clamp circuit also serves to correct for any drift or change in the amplifier. Therefore the average signal content can be changed considerably without causing a corresponding change in dc level. The sync tips are clamped to a fixed level, which is used as the dc reference.

The frequency response of the amplifier is controlled by adjusting $C2$ and the series peaking coil, $L2$. In the event dc restoration is not desired, the *dc restorer* switch, $S4$, may be opened, causing all video information to be coupled to the kinescope cathode through $C8$.

It should be noted that the input power for the first three transistors is 25 volts, while the output stage, $Q4$, has a separate collector supply input of 120 volts to allow for development of the maximum signal amplitudes necessary to drive the kinescope cathode.

Figure 8-18 schematically illustrates a monitor amplifier which employs a differential amplifier at the input to provide for rejection of *common-mode* signals. *Common-mode* refers to extraneous signals, such as hum, RF pick-up, and the like, which might be present on both the center conductor and the shield of the input cable. The amplifier shown has a bandwidth of 20 megahertz.

In the schematic, the differential amplifier consists of $Q1$ and $Q2$. Transistor $Q1$ also supplies an output for the sync separator circuitry. The differential amplifier is coupled by $C5$ to a two-stage potentiometric feedback amplifier ($Q4$ and $Q5$). Manual amplifier-gain control is accomplished by use of the *contrast* control.

The feedback amplifier is coupled through $L1$ to a cascode output stage consisting of $Q6$ and $Q7$. Transistor $Q6$ provides the composite video output signal which is direct-coupled by $L2$ to the kinescope cathode.

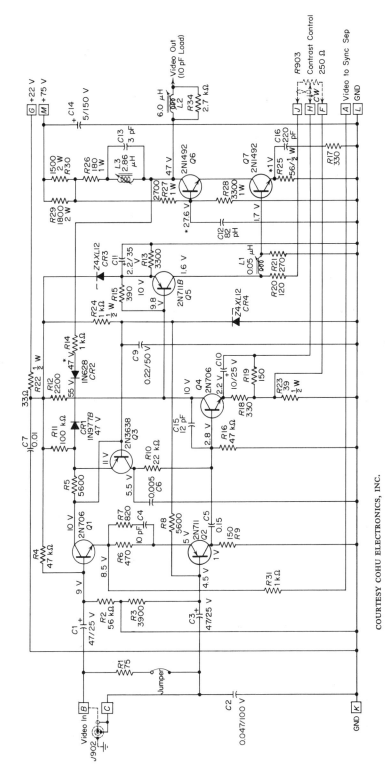

COURTESY COHU ELECTRONICS, INC.

FIGURE 8-18. A 20-MHz amplifier employing a differential input.

203

Since the dc and low-frequency components of the video signal must be restored to convey scene brightness information, restoration is accomplished by clamping the tips of the sync pulses. Clamping is accomplished by feedback from $Q6$ to $Q4$. The series elements of the feedback network are $R14$, $CR2$, Zener Diode $CR1$, grounded-base amplifiier $Q3$, and $R10$.

The sync pulses are positive at the collector of $Q6$. When the sync pulses exceed $+58$ volts, $CR1$ conducts.

# 9

## OPERATIONAL PROCEDURES FOR
## MONOCHROME TELEVISION SYSTEMS

Operational procedures for television cameras will vary with types and manufacturers of specific equipment, but the basic methods will generally be the same. While it will be found that most manufacturers recommend similar procedures for testing and adjusting their equipment, in many instances no standard method of determination exists, and the user may be hard pressed to establish the basis upon which the manufacturers' specifications were founded. For this reason, many of the examples listed here will illustrate logical and popular methods where no standard method exists.

The primary concern of this chapter will be focused on vidicon cameras, since they comprise the vast majority of existing television systems; however, many of the procedures explained (those not dealing with the vidicon tube) are representative of methods used on other types of systems as well.

### INITIAL CAMERA SETUP

When the vidicon television camera is first turned on, it may be necessary to adjust a few of the operating controls to obtain a proper picture. The three most important controls are associated with the vidicon tube itself, and are almost always labeled *beam, sensitivity,* and *focus*.

The *sensitivity* control varies the amount of voltage felt at the vidicon target electrode and, under normal conditions, will be adjusted for about

15 to 30 volts at the tube. The *beam* control is then advanced to the point where the image, as viewed on a monitor, just "wipes" on. If the *beam* control is not advanced far enough, the beam intensity will not sufficiently discharge the target, and the white areas of the presentation will appear saturated (Figure 9-1). If the *beam* control is advanced too far, the beam current will become excessive and the increase in beam spot size may cause a considerable reduction in resolution capability. Critical adjustment of the *beam* and *sensitivity* controls will result in a pleasing picture, but since there is a definite interaction between the two controls, it may be necessary to adjust each several times to obtain the desired result.

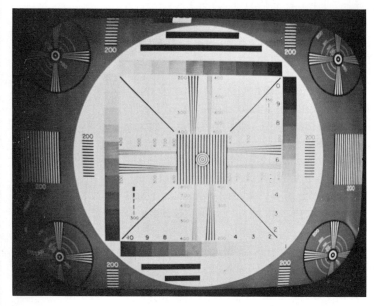

KARL KUSSEL.

FIGURE 9-1. Too little beam causes "saturated" whites.

The *focus* control that is usually provided as a front panel adjustment is, to be more precise, the *electrostatic focus* control that adjusts the focus potential within the tube itself. Generally, any controls that might be provided for the *electromagnetic focus* (i.e., control of current through the focus coil) are located within the chassis and once set, are not often disturbed. The *electrostatic focus* is adjusted until the picture appears to be the sharpest that is obtainable by use of this control. Of course, the *optical focus,* obtained by correct adjustment of the lens focus element, must also be properly set.

It is generally desirable to "drive" a vidicon at a relatively high level of sensitivity to obtain a large output signal from the tube. Such action provides for maximum output from the tube and initial preamplification stages and results in a signal-to-noise ratio that is considerably better than that which would be had if additional amplification were necessary in one of the later stages. Coupled with this, the lens iris should be opened to the diameter necessary to provide proper illumination on the tube photosensitive surface.

Once the vidicon controls have been properly positioned, the *gain* and *blanking* controls on the video processing circuitry should be adjusted to provide for a proper signal amplitude output and to maintain the correct relationship between the video and black level. An oscilloscope is almost essential in making these adjustments. The amplitude of the video signal (less sync) is commonly set to one-volt output for most closed circuit applications and about 0.7 volt for broadcast and videotape purposes. A blanking, or setup, level of 0.1 volt is usually included in this figure to assure that none of the video information is inadvertently lost by clipping at the black level (Figure 9-2). If the video signal is composite (containing sync), the synchronizing waveform is set to 0.4 volt p-p for closed circuit systems and about 0.3 volt p-p for broadcast use.

If resolution, or picture crispness, is not satisfactory, it may be necessary to adjust the aperture correction circuitry if it is provided in the system. This is usually done by the use of a switch or potentiometer control. Many

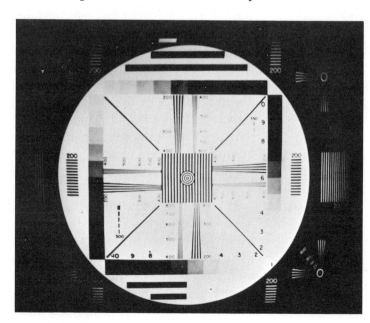

KARL KUSSEL.

FIGURE 9-2. Improperly adjusted blanking level results in clipping of black information.

types of aperture controls, in enhancing the high-frequency areas of the picture, also increase the amplitude of high-frequency noise. It is, therefore, sometimes necessary to be judicious in the use of this control, as noise may become visually disturbing in the monitor presentation. This is especially true in low light-level applications where vidicon output is low and additional gain must be introduced to maintain output signal level. The resulting increase in noise will be further boosted by the aperture circuits and may call for reduction of aperture correction to decrease the effect.

It has already been mentioned that proper adjustment of the lens of a television camera plays an important role in the initial setup procedure for the system. Improper iris and focus settings will obviously cause image deterioration. The resolution specifications of most vidicon cameras are typically somewhat poorer in the corners of the picture due to characteristics of the vidicon and scanning system, but corner resolution also suffers as the lens iris is opened to its widest positions. In fact, the entire televised picture may appear somewhat softer in its outlines when the lens iris is opened, because it is in this condition that the effects of optical aberrations are most pronounced. Depth of field is also severely reduced with wide aperture settings.

## SIGNAL-TO-NOISE MEASUREMENTS

The signal-to-noise ratio of a television system is a measurement of the ratio of video signal output to system noise, usually expressed in decibels. While no method has been officially standardized, it is common practice to employ *peak-to-peak video* voltage vs. *RMS noise* voltage to express the ratio. Thus, the peak-to-peak video signal is measured on an oscilloscope and the RMS noise should be measured with a true RMS meter, if available. If such a meter is not available, a peak-to-peak measurement of the noise can be converted to an approximate RMS noise figure by dividing it by a factor of six.

To obtain a reasonably accurate figure of merit when measuring signal-to-noise, it is first necessary to establish some ground rules. It has been mentioned elsewhere in this volume that the amount of signal output from the camera tube plays a very important role in the establishment of a signal-to-noise figure for a television system. If the tube output is low, the noise that is inherent in the input stages and succeeding portions of the amplifying system will comprise a considerable portion of the signal, and subsequent amplification will also increase this noise to the same extent that the video is amplified. Obviously then, in order to make such measurements meaningful on a comparative basis, it is necessary to establish a standard camera setup procedure which will provide for proper camera tube output.

To be truly representative, a signal-to-noise figure should be referenced to the resolution and sensitivity specifications of a television camera system. In other words, if a camera is said to produce 800 lines of horizontal resolution with 1 foot-candle of illumination on the vidicon faceplate, the signal-to-noise ratio should logically be measured under exactly these conditions. This is quite important, since the noise figure increases at low light levels when it is necessary to increase amplifier gain to compensate for reduced camera tube output. It is also often necessary to make use of aperture correction circuits to realize the maximum specified resolution of television cameras and this also increases noise levels, especially in the higher frequency areas of the system bandwidth. Consequently, immediately prior to any measurement of signal-to-noise, the test fixture lighting (or lens iris setting) should be adjusted to provide for the proper camera tube illumination. System adjustments (gain, aperture, etc.) should then be made to obtain a monitor presentation which fulfills the camera specifications, particularly in the area of resolution capabilities.

Referring to Figure 9-3, it can be seen that the light transfer characteristic of a typical vidicon (RCA type 4503) yields an output signal current of 0.2 microampere when one foot-candle of illumination is applied to the vidicon faceplate of a tube which has been adjusted for 0.02 microampere of dark current. Since the light transfer characteristics of a particular type of vidicon are generally reasonably consistent from tube to tube, obtaining 0.2 microampere of signal current with the tube set up as above will assure that the illumination on the faceplate of the tube is approximately one foot-candle. This method is generally considered to be at least as accurate as measuring the scene illumination and computing lens transmission losses, aperture inaccuracies, etc.

Using the above example, it is possible to obtain the proper target signal current, which relates to illumination, by using a microammeter to measure first the dark current of the tube and then the signal current. The procedure is to first cap the lens of the camera to assure that no signal current will be developed, and obtain a measurement of the target dark current. Target voltage should then be adjusted to obtain 0.02 microampere of dark current so that the center curve of Figure 9-3 may be used. If the camera lens is then uncovered, there will be an increase in current due to the signal that is developed by the camera tube. Subtracting the dark current from this figure will yield the actual signal current for the camera tube. The lens (or lighting) should be adjusted to provide for 0.22 microampere of current. Subtracting the dark current (0.02 microampere) from this shows the signal current to be 0.2 microampere which, according to the graph, corresponds to one foot-candle of faceplate illumination.

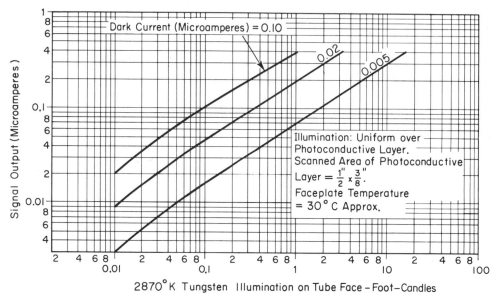

FIGURE 9-3. Light transfer characteristics of RCA 4503 vidicon.

With the television camera set up in the above manner, measuring signal-to-noise is simply a matter of comparing the video output of the system with the noise output. As previously indicated, this can be conveniently accomplished with an oscilloscope. The video output is generally set to 1-volt output amplitude, *not* counting the sync pulses, by adjustment of the *gain* control. (Once the vidicon has been set up for proper output, the *beam* and *sensitivity* controls must *not* be adjusted, or the basic reference will be lost and the resulting measurement will be meaningless.) It is also very important to note that the scene which is being used for this test should contain an area of content which will give a reference between absolute black and white. This assures the widest possible swing of vidicon output voltage, and the 1 volt of system video output will represent all tonal gradients between black and white. If it does not, it will be necessary to increase the gain of the amplifier to "stretch" the smaller voltage swing, which results from lack of scene contrast, to obtain the 1-volt output. Such action will increase the noise amplitude and results in a less satisfactory signal-to-noise ratio.

When the video output has been adjusted to the 1-volt level, the lens is then capped and all video information is thereby eliminated. If the oscilloscope is used to examine the output, the noise of the system will be present during the interval which lies between horizontal blanking pulses. A peak-to-peak reading of this noise voltage, divided by a factor of six, will yield a good approximation of its RMS value. Comparison of this value with the video signal (1 volt in this example) will show the signal-to-noise ratio of the system, which can then be converted into decibels (Figure 9-4).

While the above method is a typical means of obtaining a true signal-to-noise indication, many television manufacturers' measurements are made using a "weighting" filter. This device is generally attached to the video-output terminal of a system to lower the amplitude of the high-frequency components of the noise signal by a factor determined by its relative objectionability. It has been experimentally determined that high-frequency noise, as viewed

| | |
|---|---|
| 10:1 | 20 db |
| 20:1 | 26 db |
| 50:1 | 34 db |
| 70:1 | 37 db |
| 100:1 | 40 db |
| 120:1 | 41.6 db |
| 140:1 | 43 db |
| 160:1 | 44.1 db |
| 180:1 | 45.1 db |
| 200:1 | 46 db |
| 250:1 | 48 db |
| 300:1 | 49.8 db |

FIGURE 9-4. Typical voltage ratios expressed in decibels.

on a monitor or receiver, is not as objectionable to the human observer as is lower-frequency noise of the same amplitude and polarity. This is obviously due to the fact that low-frequency noise causes a more visible disturbance in the presentation than do higher-frequency noise elements. Thus, a filter is used which rolls off at the higher frequencies, attenuating the less objectionable elements. A properly designed filter will have a characteristic curve which approximates the relative objectionability of noise as seen by the eye. The high-frequency noise, therefore, carries less weight than its lower-frequency counterpart.

The use of a weighting filter naturally results in a somewhat higher ratio than would be had by direct measurement. Measurements which are made without the weighting filter should, therefore, be expected to produce a lower ratio than a published figure of merit for a particular system, which may have been acquired with the use of an appropriate filter.

## BANDWIDTH DETERMINATION

While it is easy to determine the approximate bandwidth characteristics of an amplifier system by using a resolution chart, it is sometimes desirable to reaffirm the exact response curve by "sweeping" the system with an appropriate signal generator.

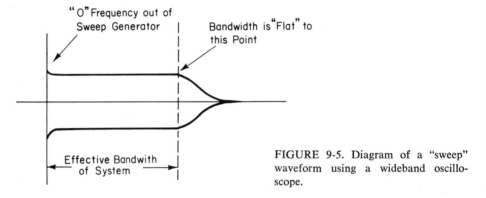

FIGURE 9-5. Diagram of a "sweep" waveform using a wideband oscilloscope.

A sweep generator varies its output frequency at a predetermined rate from some minimum frequency (usually about zero in this case) to a maximum frequency which lies beyond the upper bandwidth limits of the television system under test. The generator must, of course, maintain a constant amplitude output throughout the entire sweep excursion. If an oscilloscope is used to view the output of the system, the waveform might appear as indicated in Figure 9-5. It is very important that the oscilloscope have a bandwidth characteristic that does not attenuate the upper frequencies or the resultant waveform will not be representative of the amplifier.

Because oscilloscopes with the required bandwidth are many times not readily available, it is common practice to use a *detector probe* at the oscilloscope input to detect the envelope characteristics of the amplifier output, and present them on the scope as a line representation of the bandwidth characteristic (Figure 9-6). Also shown is the schematic of a typical detector, and an illustration of the equipment configuration when performing such tests. It is important to follow the manufacturers' instructions when performing bandwidth tests.

Since the output impedance of any particular sweep generator is not likely to be the same as that of a vidicon, it is necessary to employ an inter-

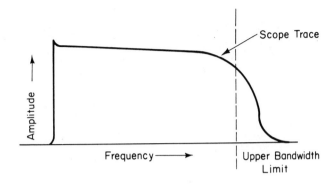

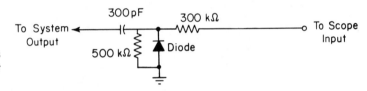

FIGURE 9-6. Detector probe allows use of low-frequency oscilloscope for bandwidth measurements.

mediate device which will simulate the characteristics of a vidicon tube. Such *vidicon simulators* will generally have output impedances and capacities that are very similar to the vidicon, and will sometimes even simulate the effective frequency response of a particular tube. Therefore, if such a device is used between the sweep generator output and the amplifier input, it will satisfy the input requirements for which the amplifier was designed.

## MEASURING LINEARITY AND GEOMETRIC DISTORTION

Figure 9-7 illustrates an equipment setup which provides a means of checking linearity and geometric distortion. The camera is viewing a standard RETMA linearity chart, thus providing a pattern on the vidicon faceplate with evenly spaced reference circles. The chart has the same 4:3 aspect ratio as the television system and, if the raster is just exactly filled with the image of the chart, measurements made with it can be quite precise. Each small circle on the chart has an inner radius of 1 per cent of the picture height, and the radius to the outside of the black portion of the circle is 2 per cent of the picture height. Total circle diameter is therefore 4 per cent of the picture height.

COURTESY COHU ELECTRONICS, INC.

FIGURE 9-7. Equipment set-up for measuring linearity.

Figure 9-7 also shows a *dot-bar generator* (also known as a *grating* generator, or *crosshatch* generator) sitting on top of the camera control unit. It generates a series of narrow pulses that are synchronized to the horizontal and vertical scanning frequencies of the television system; these may be fed directly into a monitor and will appear as narrow vertical or horizontal lines, depending upon their frequency. Figure 9-8 illustrates a monitor presentation obtained when a series of such pulses are applied at the input to the monitor. Because of the accuracy of the generator, the lines are known to be equidistant from each other with respect to time. The lines have been superimposed on the display of the chart by addition of the generator output to the camera video signal. This establishes an unalterable relationship between the generator and camera signals. Thus, any displacement of the circles in a portion of the display with respect to the generator derived lines will be due only to nonlinearity in the camera system. It is important to realize that the monitor does not affect this procedure in the least, since any distortion that the monitor may introduce is applied to both of the input signals and does not change their relationship with each other in any way.

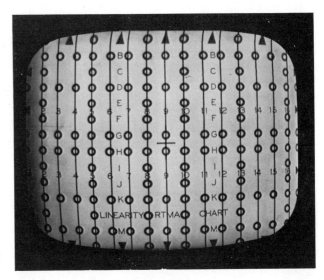

FIGURE 9-8. Vertical bars allow check on horizontal linearity.

Figure 9-9 shows the same monitor presentation with horizontal bars added to provide a reference for checking vertical linearity. These bars are coming from the same generator that provided the vertical bars. The intersection of the bars provides a convenient reference for the circles, if properly adjusted. The dot-bar generator also can generate a signal that will cause the individual lines to disappear *except* where they intersect, giving

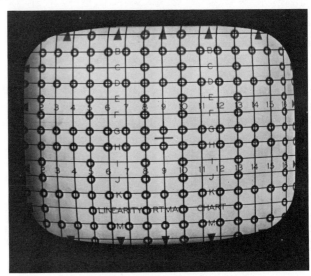

FIGURE 9-9. Cross hatch pattern increases ability to check vertical linearity.

a pattern of dots that may be used for close examination of camera linearity (Figure 9-10). If every one of these dots were to fall within the light portion of each circle on the chart, the combined linearity and geometric distortion of the camera would be within plus-or-minus 1 per cent (relative to picture height). If the dots were to be within only the outer radii, or black portion, of the circles, the figure of merit would be derated to plus-or-minus 2 per cent. Any of the dots that fall outside the circles entirely would be deviating from optimum by a factor of plus-or-minus 3 per cent or greater.

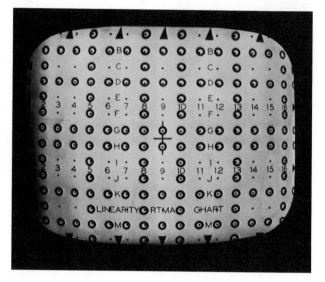

COURTESY COHU ELECTRONICS, INC.

FIGURE 9-10. Dot pattern provides critical test of overall linearity and geometric distortion.

The controls in the associated camera circuitry should be adjusted to provide the best possible linearity while observing the monitor presentation and attempting to get all of the dots as near the center of their respective circles as is possible.

Referring again to Figure 9-10, it may be noted that all of the dots in the circles that comprise the two centermost horizontal rows of circles all fall within the center portion of the circles. However, if the extreme corners of the picture are examined, it will be noticed that some of the circles are horizontally displaced so that the dots are moved slightly into the black circumference of the circles. Because the horizontal linearity (a function of scan waveforms) was shown to be within plus-or-minus 1 per cent by viewing the center rows of circles, this deviation can only be caused by some other form of distortion (yoke imperfection, etc.) and must be termed *geometric distortion*. Of course, linearity imperfections are actually a form of geometric distortion as well, but common usage separates the two for convenience of description.

*Monitor linearity* is determined by routing the output of the dot-bar generator into it, while capping the lens of the camera so that the linearity chart does not appear on the presentation. The lines that then appear on the monitor are known to be of a constant frequency, and in a perfectly linear monitor should be equally spaced, as viewed on the screen. To determine this, it is necessary to project the linearity chart onto the monitor face with a slide projector or other means, or to provide a transparent overlay of the correct size to lay against the monitor kinescope. Linearity checked in this manner eliminates the camera system from the test, except for the function of providing sync signals. Some signal generators may also provide the necessary sync signals.

In the event that a linearity chart is not available, another method of determining the linearity of a monitor would be to simply measure the spacing of the lines.

If any deviation in monitor linearity falls outside allowable limits, it will be necessary to adjust the monitor linearity circuits to bring the lines (or dots) into the circles of the overlay or projected image (or provide proper spacing if a chart is not used).

## OVERSCAN AND UNDERSCAN

A relatively simple procedure such as causing a beam to trace a raster which is larger than normal (*overscan*) or smaller than normal (*underscan*) can result in some confusion when two scanning systems are being employed, as is the case in television.

### The Monitor Raster

Regardless of monitor screen size, it is generally desirable to scan the entire area of the picture tube for presentation of the televised image. In general practice, the electron beam is deflected just slightly farther than the visible edge of the tube to eliminate the distraction that might result from the edges of the raster being visible. For general observation purposes such a scanning method is considered proper. There are times, however, when it may be necessary to reduce the monitor sweeps to be able to see all of the information being taken from the vidicon.

The monitor may normally be underscanned by reducing the width and height by adjusting the respective controls within the monitor circuitry. However, it should be noted that such action will reduce the resolution capabilities of the system, since the beam-spot size on the face of the kinescope remains constant, while reducing the area that it has to present an image. By way of example, a blunt pencil can be used to print many words on a large piece of paper, but if the paper is reduced to the size of a postage stamp, all efforts to produce the same words legibly, using the same pencil, would certainly fail. Such underscanning of the monitor is usually done for special purposes only.

Overscanning of a monitor is caused by expanding the sweeps of the kinescope. The apparent effect of such action is to cause an enlargement of the image within the confines of the tube face. Vertical overscanning causes more separation between scanning lines and, carried to extremes, can be objectionable. There is also considerable loss of picture information. Overscanning is not a general practice, beyond that necessary to remove the blanking signal from view.

### The Camera Raster

While underscanning the monitor results in a reduction of the image size, and overscanning has the effect of enlarging the image, varying the scanned area of the *camera* pickup tube has an effect on the monitor image that can be slightly confusing.

Underscanning the camera *increases* the apparent image size on the monitor presentation, and overscanning the camera results in an image of *decreased* apparent size. This can be realized when it is understood that enlarged camera rasters view more of the optical image focused on the pickup tube, yet this additional information, or increased portion of the image, must be presented in the same area of the monitor as the smaller original view. This has the effect of appearing to shrink the size of the objects being viewed.

Conversely, underscanning of the camera appears on the monitor as an apparent enlargement of the image.

Overscanning and underscanning (vertical or horizontal) are generally referenced to the standard 4:3 aspect ratio.

Underscanning of the photosensitive area of the camera tube can cause some rather serious problems. Scanning a raster on the tube for a period of time tends to desensitize the area being scanned to the extent that it is observable on a monitor if the tube is rotated within the yoke, or the raster is decentered or overscanned (Figure 9-11). The desensitized area is referred to as a *raster burn* and is usually difficult, and often impossible, to

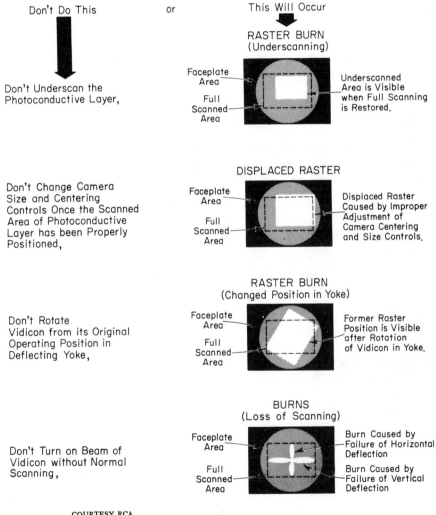

FIGURE 9-11. Scanning precautions can avoid raster burns.

remove. In the worst case, should either of the scanning sections fail, the beam would continually sweep the same straight line, imparting the same power to this small area as it had to the entire raster, and the tube would be damaged beyond repair. For this reason vidicon protection circuits are often included in cameras to protect the tube in the event of sweep failure.

## THE RESOLUTION CHART

Probably the one item most often used in the check-out and maintenance of television systems is the television resolution chart (Figure 9-12). It is designed to offer a maximum of visual indications of system perform-

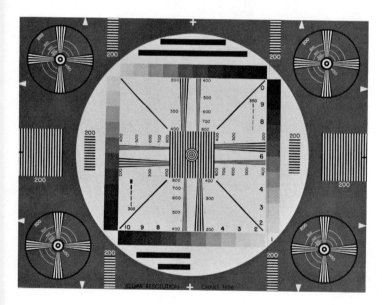

FIGURE 9-12. A RETMA resolution chart.

ance and possible malfunctions. Often the pattern is made of glass and fitted to a light box which is incorporated into a camera test fixture, as shown in Figure 9-13. This basic setup can then be used to perform many of the normal system performance tests, and provides an ideal situation for signal-to-noise measurements, etc. The chart has two sets of horizontal and vertical *resolution wedges,* which are labeled to indicate resolution elements per picture height; that is, the total number of lines (black *and* white elements each constitute a line) it would take to fill the raster height. Thus, the point labeled 400 indicates the thickness of the lines that it would take if 400 of them were used to fill the entire picture height.

It has been indicated elsewhere that horizontal resolution is a function of bandwidth. The vertical lines are used to measure *horizontal* resolution capability of a camera system. The horizontal resolution wedges are used to measure *vertical* resolution capability, which is a function of the number of scanning lines employed.

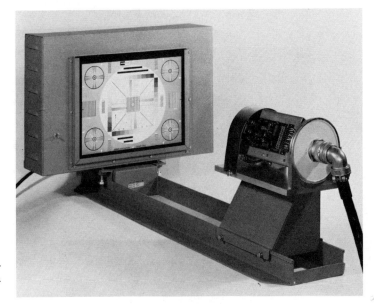

COURTESY COHU ELECTRONICS, INC.

FIGURE 9-13. A back lighted resolution chart provides critical camera alignment tests.

The test pattern also has four 10-step gray scales which may be used to test the gamma characteristics of the system. It is often a requirement, and many manufacturers specify, that television cameras be able to produce ten shades of gray to indicate a capability to provide proper tonal rendition in the televised image. A visual indication of this capability is immediately obvious when the resolution pattern is used, and use of an oscilloscope to view a section of one horizontal line of scan that passes across one of the horizontal gray scales (Figure 9-14) will reveal the linearity with which each shade of gray is amplified.

The test pattern may be used to provide an indication of system linearity. Because of the eyes' critical assessment of the linearity of a circle, any deviation from reasonable limits is immediately apparent. Also, the square groups of vertical lines at each side and the center of the pattern will provide a good cross-reference to check the horizontal linearity. Coupled to this, the square formed by the gray scale bars should form a perfect square if the aspect ratio is correctly adjusted.

The horizontal black bars located at the top and bottom of the raster, within the white area of the circle, are quite useful in setting the high-peaking circuitry of a television camera. Too much compensation, or too little, will cause a definite streaking effect to follow the transition from black to white. Such streaking can also be caused by phase shift problems within the amplifier circuits.

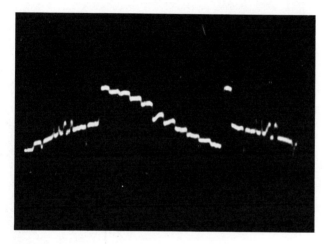

KARL KUSSEL.

FIGURE 9-14. Horizontal gray scale reveals linearity characteristics of pick-up tube and amplifiers.

Because of the inherent "flatness" of the gray background in the pattern, it makes an excellent means of checking for camera tube shading. In addition, the four diagonal black lines that radiate out from the center of the pattern are useful for checking interlace. As interlace deteriorates, the lines tend to appear jagged.

The resolution test pattern is so useful for checking camera performance that, once familiar with its use, it is quite difficult to get along without it.

## VIDEO DISTRIBUTION

Once the video signal has been developed by a camera system, it seems just a simple matter to route it to a monitor or other destination. Since most television systems use relatively short cable runs from the camera to the monitor, there is generally little difficulty. However, if transmission distances become appreciable, several problems arise.

Video transmission by use of coaxial cable is by far the most widely used method of video distribution in closed circuit applications. The cable usually matches the characteristic impedance of the electronic devices, provides for shielding during transmission, is relatively low in cost, and is easy to use. It does, however, alter the signal by attenuating the high frequencies rather severely over long cable runs. Figure 9-15 shows a chart which lists the losses encountered over various lengths of cable. This adverse effect is due largely to the capacitive action that exists between the inner conductor and outer shield. Because the two effectively form a small-value capacitor, it is easy to understand why the higher frequencies are attenuated.

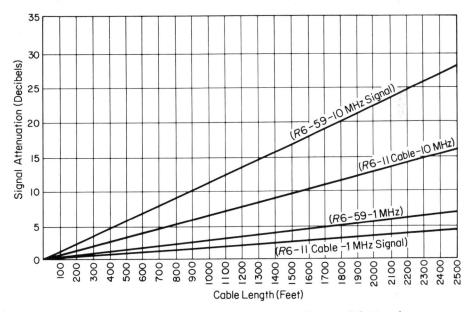

FIGURE 9-15. Cable loss determines resolution on long cable runs. RG-11 and RG-59 coax compared at 1 and 10 MHz.

When cable lengths are such that the higher-frequency elements of the video information are substantially reduced, it is common to employ "equalizing" amplifiers to process the signal and enhance the upper end of the bandwidth to restore it to its correct relative amplitude. Cable runs of unusual length may employ such amplifiers in several strategic locations along the overall length. The detailed elements of the picture may then be presented on a monitor where they would have been completely lost if equalization had not been utilized.

Any decision as to when equalization is or is not needed must be made with respect to the resolution requirements, and balanced against the costs that might be incurred. Several different types of coaxial cable are available, some of which exhibit less attenuation than others. However, it is usually true that lower loss means added cost and cable diameter is generally somewhat larger. Nevertheless, in some cases a cable with better characteristics may be more economical than an equalizing system which would otherwise be needed.

Long cable runs may also introduce a condition where ac hum is superimposed on the video signal. Another cause of hum may be the lack of a common ground between a monitor and the camera system. On long cable runs, even though a common ground may exist, the ground resistance may be such that a difference in ac potential at either end of the cable will cause a voltage drop across the coaxial shield. Since the shield is part of the signal path, interfering current in the form of hum becomes mixed with the signal as illustrated in Figure 9-16. This hum would show up as large horizontal bars of shading in the monitor presentation.

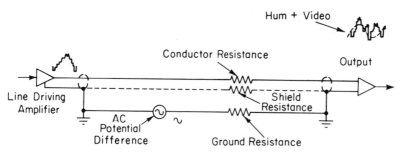

FIGURE 9-16. Long cable runs may introduce hum on the video signal.

Probably the best means of eliminating or reducing hum is to run an external grounding conductor between the two pieces of equipment. Since the requirement is to lower the ground resistance, this conductor should be as large as possible. Lowering the resistance between the two points will reduce the amount of hum that is superimposed upon the video signal, but may not completely eliminate it. If such is the case, a *clamping amplifier* may be used to eliminate the remaining ac voltage. It does this by using the horizontal sync pulses which constitute a part of the video signal to establish a constant voltage reference point. Since this occurs at a rate of 15,750 hertz (in a 525-line system) the reference will be re-established over 260 times during one cycle of the interfering 60-hertz hum signal.

Another means of eliminating hum is to use a *differential amplifier* to eliminate the disturbance. Figure 9-17 shows a simplified schematic of such a device. The hum which is present is generally "common mode" or equipresent on both the shield and center conductor of the coaxial cable and the polarity and amplitude are both the same. Thus, due to the differential action of the circuit, all of the common-mode signals are removed. Only the video signal that exists on the center conductor is passed, thus removing the unwanted hum.

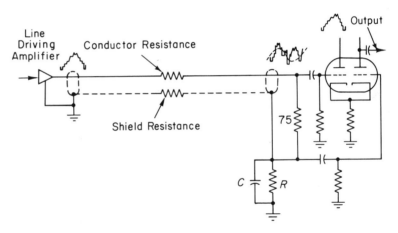

COURTESY DYNAIR ELECTRONICS, INC.

FIGURE 9-17. A differential amplifier eliminates hum.

The subject of video distribution is a complex one, worthy of a volume in itself, but the above examples are some of the more common problems encountered and are listed primarily to create an awareness of their existence. The best television camera system in the world is of little use if the distribution methods are inadequate. However, it should not be assumed that additional distribution equipment will always be required. Uncompensated cable runs of approximately a thousand feet are common and the results are generally quite good.

# PRINCIPLES OF
# COLOR TELEVISION

Color perception is very important to the human eye. It is not fully understood just how the eye perceives color, but its ability to do so certainly makes life a more interesting and enjoyable experience. Color exerts a great influence upon our everyday lives. It attracts our attention, increases perception, and even affects our emotions.

While the exact mechanics of the eye's ability to discern color are not known, it has been established that the sensation of color is directly related to the frequency of the light that is being observed. As the electromagnetic waves that constitute light change in frequency, there is a corresponding change in the color that the eye sees. One particular frequency will cause an observer to see blue, while a different frequency will give the sensation of seeing green.

Sunlight, which contains all of the frequencies in the visible spectrum, is considered to be white light. It is composed of a veritable jumble of electromagnetic frequencies, seemingly without order, yet its constituent parts form the colors that are seen in everyday life. White light may be separated into its various components by the use of a simple prism, as shown in Figure 10-1. The refractive action of the glass causes the higher frequencies to be delayed more than the lower frequencies. The light rays therefore exit the prism at angles which differ according to each individual frequency present. This essentially files the frequencies in a sequential order according to their wavelength. The result, when projected onto a flat white surface, yields the various colors that are contained in the original white light. It can be determined that the red color is the lowest in frequency since it was refracted, or bent, the least. Orange, yellow, green, blue, and violet are arranged in the order of ascending frequency that constitutes each color. There are no sharply defined borders between colors. The effect is that of one color blending into the other and it becomes a matter of individual interpretation where one predominate color ends and another begins.

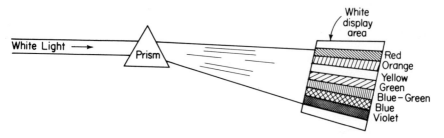

FIGURE 10-1. Refractive action of a prism separates light into its constituent components.

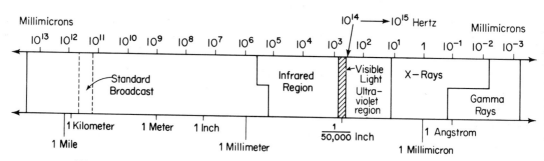

FIGURE 10-2. The relative position of visible light in the electromagnetic spectrum.

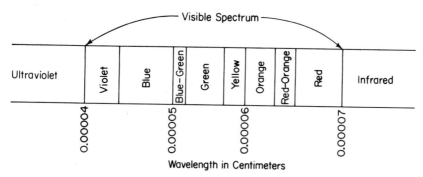

FIGURE 10.3. The visible spectrum expressed in centimeters.

Figure 10-2 shows the visible portion of the spectrum, and its relative position in the total electromagnetic spectrum is detailed in Figure 10-3. It can be defined as that portion of the spectrum that stimulates the eye and causes the sensation of seeing light. The color of the light that one sees depends upon the wavelengths that reach the eye.

### Measuring the Frequencies of Light

The frequencies of electromagnetic radiation that compose visible light are of an extremely high order. While it is common to refer to ordinary radio waves in terms of meters when expressing the wavelength, when light is considered it is necessary to express the wavelengths in small fractions of a centimeter. By way of example, those frequencies that are generally interpreted as red have wavelengths that are between 0.000075 and 0.000065 centimeter in length. Expressing the wavelengths of color in this manner gives a familiar unit of comparison, but it is generally not convenient to use such lengthy expressions either verbally or mathematically. For this reason, a system of measurement has been adopted that uses the *angstrom* and the *millimicron* as the basic units for measuring the wavelengths of light. Specifically:

| | |
|---|---|
| 1 micron | 1/1,000,000 meter |
| 1 millimicron | 1/1000 micron |
| 1 angstrom | 1/10,000 micron |

In terms of centimeters:

| | |
|---|---|
| 1 millimicron | $10^{-7}$ centimeter |
| 1 angstrom | $10^{-8}$ centimeter |

Figure 10-4 again illustrates the visible spectrum and defines the wavelengths of the individual color bands in terms of millimicrons.

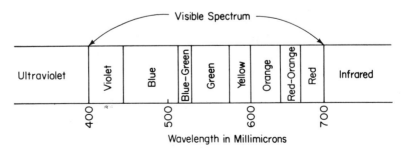

FIGURE 10-4. The visible spectrum expressed in millimicrons.

The colors that are seen in nature are not pure in the sense that they are composed of a single frequency. Reference to the preceding illustrations shows that colors such as red, green, and blue are defined as *bands* of frequencies. To illustrate, light from a red light bulb may be composed of many frequencies, but it appears red to the eye simply because *most* of the predominant frequencies emitted lie within the red portion of the spectrum. This is true for most of the colors that are normally considered as pure by the average observer. A truly pure color would consist of a single frequency, giving the light a *monochromatic* characteristic. A few light sources, such as the sodium arc lamp, can be thought of as a close approximation to a monochromatic light source because the emissions are confined to a very narrow band of frequencies. However, such conditions are rare and colors can generally be thought of as *mixtures* of frequencies.

**Reflected Color**

It has been stated that sunlight contains all of the frequencies within the visible spectrum. What then causes the eye to see different objects as different colors? It can be said that an object reflects that portion of sunlight that is its characteristic color and absorbs all other frequencies. A green leaf, therefore, simply reflects those frequencies that correspond to the green coloring and absorbs all other frequencies in the visible spectrum. Complete absorption of the other frequencies is never fully achieved, but the predominant reflection is in the green region and this is the color that is seen by the human eye. Similarly, a red object reflects red frequencies and absorbs all others. Thus, the color of an object is defined by the frequencies that it reflects.

It is important to note that an object will not appear as its characteristic color if that color is not present in the light emitted from the illuminating source. If a red light is used to illuminate a green object, the object will appear black because no green components are present in the light and the red is absorbed almost in its entirety. Blue light directed at a red object will result in the same effect. Reflected color is thus dependent upon the illumination components as well as the reflection characteristics of the object.

For an object to appear white, it must be able to reflect all colors equally well. This seems obvious enough when it is remembered that white screens are used for the display of color slides and movies, but it takes on added significance when the various color frequencies are considered. Since white light is understood to be composed of various frequencies, it follows that the reflection must contain them all to make the object appear truly white. Conversely, an object appears black because it absorbs all of the frequencies contained within the visible spectrum and very little reflection takes place.

### Color Separation by Filtering

The incandescent bulb in a flashlight emits white light, but if a piece of red glass is used to cover the reflector, the resulting light beam will be red. This is caused by the *filtering* action of the colored glass, and is somewhat analogous to color reflection. The red filter allows those frequencies that lie in the red portion of the spectrum to pass and absorbs all others. Color motion picture film and slide transparencies make good use of this effect. The many color elements in each projected image are the result of the filtering action of the colored film emulsion upon the original white light of the projector.

Figure 10-5 is a graphical representation of the frequencies that might pass through a typical filter. There is generally no sharp cutoff point at

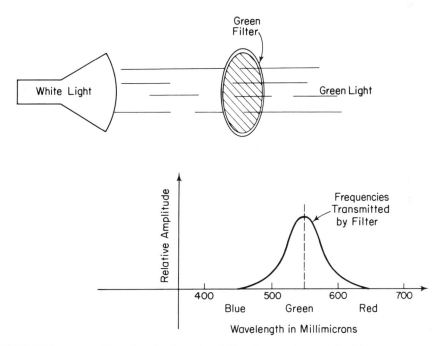

FIGURE 10-5. A green filter absorbs the red and blue frequencies contained in the original white light.

which the filter begins to absorb frequencies rather than allowing them to pass. The filter illustrated was chosen to allow the passage of green frequencies, but it can be seen that frequencies which correspond to red and blue are also present in lesser amounts. Green would be the predominant color, however, and the eye would interpret it as such.

Both the reflective action of colored objects and the selective transmission of color filters rely upon the fact that the resultant color is achieved by the *absorption* of all frequencies except those of the characteristic color. In effect, the unwanted frequencies are subtracted from the white light. For this reason, color generation using these principles is called the *subtractive* method.

### Color by Addition

If two circles of colored light are projected upon a white screen, as illustrated in Figure 10-6, the reflected light faithfully reproduces the original colors. If the two circles are allowed to overlap, the red and the green light combine and the reflection appears yellow. The particular shade of yellow depends upon the characteristic wavelengths of the red and green light and the relative intensities of the two. If the red light is made more intense than the green, the yellow area will tend to appear orange in color. If the green is more intense, the yellow will become a yellow-green color. To understand why this occurs, it is necessary to examine the color response of the human eye.

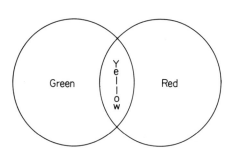

FIGURE 10-6. Projected red and green light combines to form yellow.

It is thought that there are three separate sensory systems within the eye which respond individually to red, green, and blue. Figure 10-7 is a graphic representation of the relative amount of the three primary colors necessary to stimulate the three receptor systems to achieve a full gamut of colors. It may be noted that the band of frequencies to which each sensory system responds is quite broad, with considerable overlapping involved, especially in the red and green frequencies. The green sensors, for example, show a response to the red and blue frequencies as well as green, although their sensitivity is considerably reduced in these areas. When a color such as yellow is observed, it excites *both* the red and green sensors in such a manner that the brain interprets the resultant sensation as being yellow. Similarly, a blue-green (or cyan) color, whose frequency lies between the green and blue portions of the spectrum, will cause simultaneous excitation of both the blue and green sensors and the cyan color will be seen. White light excites all three sensory systems to the extent that the result is interpreted as being white.

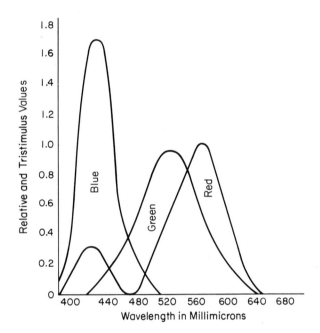

FIGURE 10-7. Color mixture curves using the three standard TV primary colors to achieve the full range of visible colors.

Referring again to the color circles in Figure 10-6, it seems apparent that there are no frequencies present that would constitute yellow. The sensation of yellow is caused because the red and green frequencies, when focused on the same area of the retina of the eye, excite the red and green sensors to the same degree that they would be affected by yellow light. Since the stimulation of the sensor nerves is the same in both cases, the resultant color that is seen appears the same.

It may have occurred to the reader that the color purple occurs nowhere on the spectrum chart. This is an example of the eye seeing a color that is not represented by any single frequency. Purple is a combination of red and blue light. These two colors lie on opposite ends of the visible spectrum, yet, when combined, the eye sees a color that seems as pure as any color represented by a single frequency. The simultaneous excitation of the red and blue sensors is responsible for the purple color.

Figure 10-8A illustrates the effects of three colored circles of light projected upon a white screen. The red, blue, and green are chosen because, when combined in the proper proportions, it is possible to produce a wider variety of colors than would be possible by mixture of any other three colors. This is undoubtedly due to the peculiar sensitivity of the human eye in these three regions. For this reason, red, green, and blue are termed *primary colors* and are the foundation upon which color television is built.

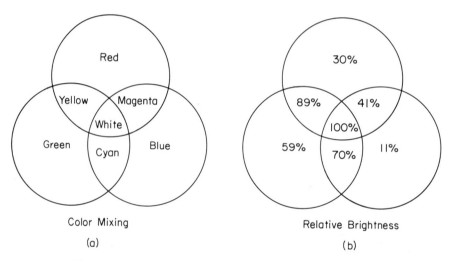

FIGURE 10-8. Red, blue, and green primaries yield colors that vary in relative brightness.

Attention to the diagram will illustrate that white light may be obtained by the proper mixture of the three primaries. However, it is important to point out that the relative brightness of the three colors, when viewed separately, will not appear equal. When white light is produced by equal amounts of the three primaries, and the three of them are individually examined, the green will appear to be the brightest because the eye has its greatest sensitivity in this area. The red will appear to be about one-half as bright as the green, and the blue will seem approximately one-third the brightness of the red or one-sixth as bright as the green. These facts have a great bearing upon the composition of a television color signal, as will presently be seen. The actual values of relative brightness that have been adopted for use in television, using white as 100 per cent brightness, are:

| | |
|---|---|
| Green | 59% |
| Red | 30% |
| Blue | 11% |

Figure 10-8B shows that the brightness of the colors produced at the junction of any two of the primaries is the result of direct addition of the brightness of the two primaries. Yellow, for example, appears to be 89 per cent as bright as the reference white, reflecting the addition of the 59 per cent brightness of green and the 30 per cent brightness of red.

### Hue, Saturation, and Brightness

The words *hue, saturation,* and *brightness* undoubtedly convey meaning to the average person because of their general usage in everyday language. However, the mental images that arise in connection with these words are not always accurate.

The term *hue* is used to describe a particular color. Although it is associated with the frequency of light, it is difficult to describe it as such since many colors are composed of frequency elements from different portions of the spectrum. The word is generally not used when referring to black and white or the shades of gray which lie between the two, probably because common usage does not ascribe color to these areas. The term "red" describes hue, as does "yellow," "green," and "violet." However, words such as "red" are not entirely descriptive, as there may be literally hundreds of shades of red which are discernible to the eye because they each possess a characteristically different hue. For example, if a small amount of blue light is added to an intense red light, the result will still appear predominantly red, but there will have been a detectable change in hue.

The term *saturation* can probably best be described by the use of an illustration. Consider two projectors directed at a viewing screen, one emitting white light and the other green light. If the two are superimposed, the green will be visible, but it will appear to be diluted with the white light. If the intensity of the white and green projectors is equal, it may be said that the green is 50 per cent saturated. If the output of the white projector is slowly lowered in intensity, the screen will appear greener in nature as the amount of white light is reduced. With no white light diluting the purity of the green, it can be said that the green is 100 per cent saturated. It is important to note that the frequency content of the green source was not changed, and its intensity remained unaltered. Nevertheless, saturation is generally expressed with relation to hue, giving rise to such expressions as light blue, medium blue, and dark blue. However, this should not be construed to mean that the basic blue frequency elements have been altered, even though the same terms can be used to describe an actual change in color frequency, or hue.

The *brightness,* or intensity, of a particular color may be varied without affecting either the hue or saturation. If the green projector mentioned previously were the sole source of screen illumination, increasing its output would result in an increase in the color brightness. The hue and saturation would not change, although there are certain characteristics of the human eye that might make it appear so.

### Resolving Color Elements

Color that is easily discernible on a large object becomes less so as the object size is decreased. A ball that is obviously red when viewed at close range appears to be gray or black at some critical distance, due to its smaller apparent size. Similar results are obtained with other colors although the critical object size differs with different colors. Below a certain size, and somewhat dependent upon ambient conditions, the eye loses its ability to distinguish color of any sort, and objects appear as sources of luminance information only. That is, they are interpreted as black, white, or various shades of gray, and no color information is conveyed.

The loss of color vision with a decrease in size is most pronounced for objects that are blue. When the critical angle of view is approached, it becomes difficult to distinguish the difference between a blue object and another object of similar size that is gray. Other objects of this size that might be colored red or green would still be quite visible as individual colors, but if unit size were further reduced the red would begin to appear gray also. Finally, the green would shift to an apparent gray as the critical visual angle for that color was achieved. All three of the original colored objects would still be visible as elements of some unit brightness, depending upon their original brightness, but all color information would be lost to the eye.

The loss of color vision related to objects of small size has an important bearing upon the concepts of color television. Since resolution elements below a certain size cannot be discerned as color, it is not necessary to provide wide bandwidths for the signals containing color information.

The small elements of the picture will be interpreted as black and white (or monochrome) information in any event, so it seems logical to transmit them as such. The high resolution elements of the televised picture are therefore transmitted as luminance information only, and the larger, lower-frequency elements contain color information.

Another important characteristic of the human eye is its tendency to integrate small units of color that are in close proximity to one another. Figure 10-9 shows three dots equally spaced in a triangular configuration. If the red, blue, and green dots are assumed to be sources of equal intensity, each color will be clearly visible at close range.

However, if the group is moved to a point some distance away, the eye will tend to view the three elements as a single point and the three individual colors will blend together. The integrating action of the eye will therefore cause the objects to appear as an illumination source whose color is determined by the relative intensities of the three individual dots. If only the red and green dots are used, the eye will interpret the result as being yellow. Similarly, use

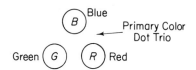

FIGURE 10-9. Small areas of color are integrated by the eye as a single resultant color depending upon the relative amounts of the original colors.

of the green and blue dots will yield a sensation of seeing the color cyan. Proper intensities of all three colors will result in a white presentation. Clearly, the principles of additive color apply here. Various other combinations of the three primary colors would have exactly the same effect as was obtained by mixing colored beams of light on a projector screen.

Close examination of the picture tube on a modern color television receiver or monitor will reveal that the phosphor surface is composed of literally hundreds of thousands of red, green, and blue dot trios similar to that shown in Figure 10-9. Proper excitation of these phosphor dots results in the wide gamut of producible colors that constitute what has been called an "added dimension" in television viewing.

## THE TELEVISION SYSTEM

Figure 10-10 is an illustration of a simple color television system. Essentially, the color camera is composed of three monochrome camera systems that receive scene information that has been selectively filtered to provide each camera tube with light from a different portion of the color spectrum.

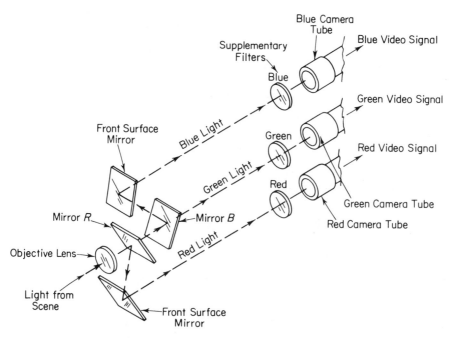

FIGURE 10-10. Optics in a color camera separate the scene into red, green, and blue elements.

The light, which is processed by the objective lens, is directed at an arrangement of mirrors which separates the light into three elements. The mirrors are coated with a special dichroic material that reflects a specific color and allows other frequencies to pass through it. Mirror *R* reflects red light and allows green and blue to pass. The red light is reflected to a front surface mirror which directs it to the red camera tube. When the light that passed through mirror *R* arrives at mirror *B,* the blue components are reflected and the green light that remains is allowed to pass through to the green camera tube. The reflected blue light arrives at the blue camera tube via its reflective mirror.

The original multicolored image that was formed by the objective lens has now been divided into its red, blue, and green elements and focused upon the indicated camera tubes. The video signals that are developed in the individual tubes are, therefore, representative of the color content of the scene. If the objective lens was focused upon an object that was entirely green, no reflection would occur at mirrors *R* and *B* and the red and blue tubes would provide no output signal. The entire signal would be developed by the green tube. On the other hand, a purple image, that contains both red and blue light, would be routed to the red and blue tubes while the green tube would not receive an image.

Scanning of the camera tubes is accomplished simultaneously, by a master deflection oscillator that drives the beams in all three tubes. This assures that the beam in each tube will be discharging the same point on the image at exactly the same instant.

The three video signals developed in the color camera represent the red, green, and blue colors of the three-primary system of colorimetry previously discussed. It is therefore possible to reproduce almost all of the colors in the visible spectrum by the selective use of these three signals. In most television sets currently in use, this is accomplished by displaying the three signals on a tricolor picture tube.

The tricolor picture tube (see Chapter Fourteen) differs from the standard monochrome picture tube in that it generates three separate electron beams to scan the phosphor surface that coats the rear side of the faceplate. The phosphor surface is also quite special, as it is composed of hundreds of thousands of small phosphor dots arranged in triangular groups across the scanned area. The individual dots in each trio emit different characteristic colors when excited by an electron beam, namely red, green, and blue.

Because the dots are small, and in close proximity to each other, they cannot be individually discerned when viewed at a moderate distance. If only the red beam is present, only the red dots will be illuminated and the entire screen will appear red during normal scanning. Increasing the beam intensity will increase the brightness of the red color. If the green beam is present, and the red and blue absent, the screen color would be green. A blue beam would similarly yield a blue color. When more than one color of phosphor dots is illuminated, the eye integrates the result and a product color is seen. As previously discussed, the color obtained is dependent upon which of the three primary colors is present, and on their relative intensities. The principle of additive color mixing applies. A white presentation may be obtained by proper adjustment of the three beams to achieve the correct intensities.

Referring to Figure 10-11, when the color camera views a yellow object, the red and green camera tubes receive the image and the blue tube remains dark. The subsequent red and green signals are amplified and sent to the color monitor. Here the red signal modulates the red beam and the green signal modulates the green beam. Lack of signal in the blue channel keeps the blue beam cut off. The red and green beams illuminate all of the adjacent red and green phosphor dots and the product color that is seen by the eye is yellow.

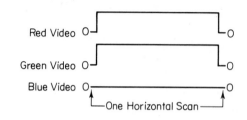

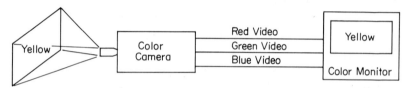

FIGURE 10-11. A yellow object is relayed to the monitor as red and green video signals.

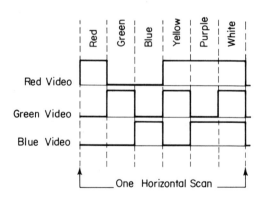

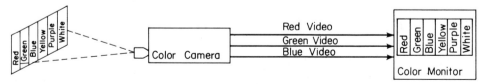

FIGURE 10-12. Multihued object generates various combinations of red, green, and blue video waveforms.

The multicolored object in Figure 10-12 will generate the video wave-forms indicated during one horizontal scanning interval. The red bar on the pattern causes a video signal to occur in the red channel while the other two channels have no output. The green and blue bars cause a similar reaction in the green and blue channels as the scan progresses. Each of these signals excites the proper beam in the color monitor and the red, green, and blue bars are produced. The yellow bar causes outputs on both the red and green channels and the excitation of the red and green beams in the monitor results in the presentation of the yellow bar. The purple bar creates video on the red and blue channels, and the white bar generates an output on all three channels.

The generation of color television images is accomplished by the sub-tractive method of color processing in the camera and the principles of additive color generation in the monitor. The range of colors that may be produced is limited primarily by the three primary colors that are used and by the capabilities of the phosphors within the color picture tube. The results that are obtained with the color television systems of today indicate that the limitations are quite minimal as interpreted by the human eye.

## THE COMPOSITE COLOR SIGNAL

For a color television system to be useful for broadcast purposes, and to increase utility in closed circuit usage, some means must be used to combine all video and color information into a single, composite signal. There is the further qualification that the signal must also be fully com-patible with existing monochrome television receivers. That is, the com-posite color signal must be generated in a manner that will allow its use in a standard monochrome receiver, producing a black and white presenta-tion with minimum disturbance from the color information. There is also the problem of containing the signals comprising the color information within the confines of the relatively narrow bandwidths allocated for com-mercial television transmission.

The compatible system of color television that is used in the United States is a result of the combined efforts of a group of engineers from many areas of the television industry. Together they formed the National Tele-vision Systems Committee (NTSC) and assumed the task of formulating a system of color transmission that would be compatible with the millions of monochrome receivers then in use. Considering the complexities of the problem, to say that they succeeded seems a gross understatement.

The NTSC color television signal is composed of two elements, the *luminance signal* and the *color signal*. The luminance signal is also called the *brightness signal* and the color signal is generally referred to as the *chrominance signal*.

### The Luminance Signal

The luminance signal obtained from a three-tube color camera system, is composed of signal elements taken from the red, green, and blue channels. This is necessary in order to convey a *total* brightness signal, because the filtering in the optical system separates the brightness elements according to their color. Because of the peculiar sensitivity of the human eye to the three primary colors, the relative amplitudes of the three camera signals are altered prior to being combined for luminance information.

Earlier in this chapter it was demonstrated that three beams of red, green, and blue light of equal intensity will produce white light when superimposed, yet when examined separately the green appears only 59 per cent as bright as the original white, the red 30 per cent, and the blue a mere 11 per cent. For this reason the luminance signal is formed using these same percentages of the three respective camera signals. That is, 59 per cent of the green signal, 30 per cent of the red signal, and 11 per cent of the blue signal are added together to obtain the total luminance signal. Thus:

$$Y = 0.59G + 0.30R + 0.11B$$

Where $Y$ = luminance signal, and $G$, $R$, and $B$ represent the green, red, and blue color signals (because of the $Y$ notation in this formula, the luminance signal is often referred to as the $Y$ signal).

### The Chrominance Signal

The chrominance signal is that portion of the composite color signal that represents the color information contained in the televised scene. It has the task of conveying the color signals developed by the three individual color channels in the camera system. Obviously, it is not possible to add the three camera signals to the luminance signal directly. To do so would simply yield an amplitude-modulated signal that would be interpreted as distorted luminance information. Therefore, it was decided to provide a subcarrier frequency that could be altered in phase and amplitude to convey color information.

The choice of a subcarrier frequency had to be such that it would not produce noticeable interference with the luminance signal when displayed on a monochrome receiver, yet, since the entire allocated bandwidth was seemingly being used by the luminance signal and sound subcarrier, it obviously had to be positioned somewhere below the four-megahertz limit on the bandwidth curve. Thus, a frequency of 3.579545 megahertz was chosen as the color subcarrier frequency. True, this would be noticeable on a monochrome receiver, but the frequency was high enough so that any variation in luminance that resulted would be seen as a very small resolution element in the picture and would not be as visually disturbing as that obtained when using a lower frequency. Still, the effects would be noticeable.

### Reducing Subcarrier Interference

The unusual subcarrier frequency takes on added significance when considering subcarrier interference as seen on the television picture tube. The subcarrier frequency was selected to be an odd multiple of one-half the horizontal scanning frequency of the television system. To illustrate the reason for this, it is convenient to assume that a simple five-line scanning system is being used.

Figure 10-13 illustrates the effects of a sinewave luminance signal that is an odd harmonic of one-half of the scanning frequency. The positive excursions of the signal present a unit area of brightness on the picture tube and the negative-going part of the signal represents dark areas on the presentation. The sinewave completes 2½ cycles during one active horizontal interval. In the illustration this represents three dark areas and

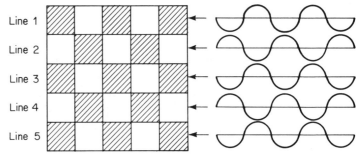

FIGURE 10-13. A sinewave luminance signal that is an odd harmonic of one-half the scanning frequency generates a checkerboard pattern.

Line 1

Line 2

Line 3

Line 4

Line 5

two areas of brightness. Because of the extra half-cycle, the next horizontal scan begins with an area of brightness and the entire line contains only two dark areas. It may be noted that the bright areas that occur during line two are horizontally coincident with the dark areas of line one. This same off-set occurs on each succeeding scanning line, causing a checkerboard pattern to be obtained.

Because the scanning rate in the example utilizes an odd number of horizontal scans for each completed presentation, the luminance signal will be 180 degrees out-of-phase with the previous signal as line number one is again scanned. Thus, the black areas that existed during the first scan will be replaced by bright areas and the original bright areas will be dark, and as the scanning process continues, the checkerboard pattern obtained on the second scan will be the reverse of that which was generated originally.

The total effect of the above process on the human eye is one of cancellation. If the process is repeated fast enough, the persistency of the eye will blend the patterns together and their visibility will be considerably reduced. The color subcarrier frequency was chosen to produce just such an effect, and it consequently goes unnoticed when viewed on a monochrome receiver.

### Frequency Interleaving

Another very important factor that had to be considered in the choice of a color subcarrier frequency was the disturbances that would be caused by placing it in an area already occupied by frequencies generated by the luminance signal. By using a process known as *frequency interleaving* it is possible to contain both of these signals within the same bandwidth.

Frequency interleaving in television transmission is possible because of the relationship of the video signal to the scanning frequencies that are used to develop it. The scanning frequencies occur at a constant rate of 15,750 hertz for the horizontal and 60 hertz for the vertical. It has been determined that the energy content of the video signal is contained in individual energy "bundles" that occur at harmonics of the horizontal and vertical scanning frequencies, shown in Figure 10-14. The shape of each energy bundle shows a peak at exact harmonics of the horizontal scanning frequency. The lower amplitude excursions that occur immediately to each side are spaced at 60-hertz intervals and represent harmonics of the vertical scanning rate. The vertical sidebands contain less energy than the horizontal because of the lower rate of vertical scanning repetition.

The above description assumes that all video information will occur at exact multiples of the scanning frequency, and it can be shown that the video may indeed be represented as being composed of a large number of pure sinewaves which are harmonics of the scanning frequencies. Thus, if one horizontal line of video signal were to be separated into its individual sinewave components, each would be an exact harmonic of the line rate.

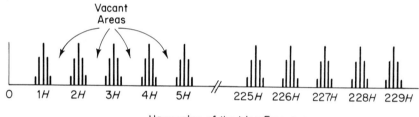

FIGURE 10-14. Video harmonics occur at multiples of the line frequency.

It may be seen from the above that a good share of the bandwidth in a monochrome television signal goes unused because of the spacing between the energy bundles. Thus, it would seem that more information could be relayed within the bandwidth limits if it could be made to occupy the areas that lie between the established harmonic groups.

It has already been established that the subcarrier frequency must be an odd multiple of one-half the horizontal line frequency to produce cancellation of its luminance product in monochrome receivers. It should be noted that the unused energy areas shown in Figure 10-14 occur at frequencies which are also odd multiples of one-half the horizontal line frequency. Therefore, since the chrominance information in a color video signal is generated by the same scanning frequencies that produce the luminance information, and will have essentially the same energy distribution, it may logically occupy the unused areas.

Figure 10-15 illustrates the energy content of a color television signal using a color subcarrier frequenciy that is an odd multiple of one-half the line frequency.

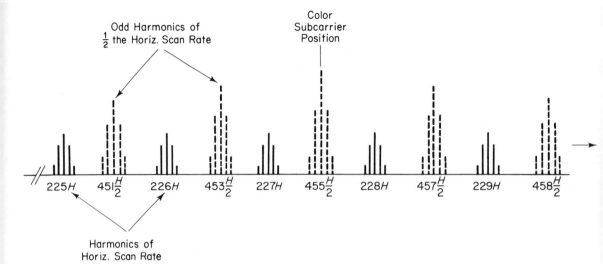

FIGURE 10-15. Color subcarrier generates harmonics that are odd multiples of one-half the line frequency.

### Subcarrier Frequency Considerations

The modulated color signal may have sidebands that extend above the subcarrier frequency some 0.6 megahertz, so it follows that the subcarrier must be positioned at least 0.6 megahertz below the upper frequency limit of the video bandwidth of the color television receiver. Since this is generally about 4.2 megahertz, subtracting 0.6 megahertz yields a subcarrier frequency of approximately 3.6 megahertz. However, it has been indicated that the frequency chosen must be an odd multiple of one-half the horizontal line frequency. If we choose the 455th harmonic of one-half the line frequency, we obtain:

$$7875 \times 455 = 3,583,125 \text{ Hz}$$

or approximately 3.58 megahertz.

This certainly satisfies all the prerequisites that have been set down as parameters for the subcarrier. It is relatively high in frequency, is an odd multiple of one-half of the line rate, and is at least 0.6 megahertz below the upper bandwidth limit. However, because a subcarrier of this frequency produced an objectionable beat frequency with the 4.5-megahertz sound carrier, the present subcarrier of 3.579545 megahertz was chosen.

The new frequency limits the beat patterns that were produced by the previous carrier. Of course, lowering the frequency resulted in a sub-carrier that was no longer an odd multiple of one-half the line frequency. Nevertheless, this requirement could again be satisfied if one-half the line frequency were lowered to the 455th submultiple of the new subcarrier frequency, or:

$$3.579545 \text{ MHz} \div 455 = 7867.132 \text{ Hz}$$

This would make the horizontal line frequency:

$$7867.132 \times 2 = 15,734.264 \text{ Hz}$$

This horizontal frequency poses no problem with existing mono-chrome receivers because the circuitry involved can easily adjust to such a small frequency change.

The vertical frequency becomes:

$$15,734.264 \div 262\frac{1}{2} = 59.94 \text{ Hz}$$

which is much less than one cycle difference from the original 60 hertz, so again, no circuit changes are necessary and these new frequencies are completely compatible with previously existing receivers.

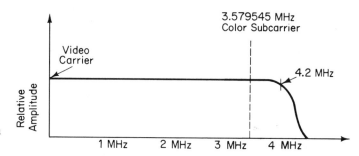

FIGURE 10-16. Subcarrier position on bandwidth curve.

From the foregoing discussions it can be appreciated that the establishment of a color subcarrier frequency involved a large number of considerations. Figure 10-16 shows the final position of the subcarrier within the bandwidth curve of the standard video spectrum. It has already been stated that the upper sideband limitation of the color signal is limited to about 0.6 megahertz. The lower sideband is allowed to extend below the subcarrier frequency by 1.5 megahertz.

The chrominance, or color, information of the NTSC color television signal conveys the hue and saturation elements of the televised scene. Hue is transmitted as a variation in phase of the subcarrier frequency, and saturation is expressed as amplitude modulation of the subcarrier. However, the three color video signals that were obtained from the television camera must be considerably altered before any modulation of the subcarrier takes place.

### The R-Y and B-Y Signals

It has been previously stated that the $Y$, or luminance, signal of the NTSC color system is essentially composed of $0.59G$ plus $0.30R$ plus $0.11B$. This arrangement is achieved by *matrixing* the individual outputs of the three video channels of the color camera system. Figure 10-17 illustrates a method by which this may be done.

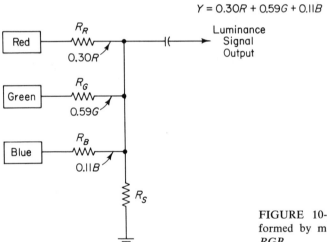

$$Y = 0.30R + 0.59G + 0.11B$$

FIGURE 10-17. Luminance signal is formed by matrixing proper values of *RGB*.

If output is present from the red channel only, and an initial potential of 1 volt is assumed, the voltage dropping action of $R_r$, in conjunction with the signal developing resistor $R_s$, will result in an output of 0.30 volt. If only the green channel is providing an output, the action of $R_g$ and $R_s$ will provide an output which is 0.59 volt in amplitude, again assuming an original 1 volt present at the matrix input. In the case of the blue channel, the 1-volt input to $R_b$ would be reduced to 0.11 volt at the output. If a 1-volt signal were present at all three channels, the dropping resistors $R_r$, $R_g$, and $R_b$ would perform their voltage reducing functions as before, and all three would be summed by the total current flow through $R_s$. Thus, the matrix will provide a luminance-signal output that is 1 volt in amplitude, and satisfies the luminance equation previously derived:

$$Y = 0.30R + 0.59G + 0.11B$$

If $Y$ is inverted in polarity and subtracted from each of the individual color signals, the following equations are obtained:

$$R - Y = R - (0.30R + 0.59G + 0.11B) = \phantom{-}0.70R - 0.59G - 0.11B$$
$$G - Y = G - (0.30R + 0.59G + 0.11B) = -0.30R + 0.41G - 0.11B$$
$$B - Y = B - (0.30R + 0.59G + 0.11B) = -0.30R - 0.59G + 0.89B$$

Subtracting the $Y$ component in this manner removes all luminance information and leaves three signals that represent only the color information.

It is not necessary to use all three of these signals, since any one of them can be obtained by proper manipulation of the other two and the $Y$ signal. For instance, if only the $R$-$Y$ and $B$-$Y$ signals are used, the $G$-$Y$ may be obtained by multiplying $R$-$Y$ by 0.51 and $B$-$Y$ by 0.19, and then adding the two resultants together. If some method of doing this is utilized in the monitor or receiver, it is not necessary to transmit the $G$-$Y$ signal. This is exactly what is done in the NTSC color system.

Figure 10-18 is a simplified illustration of the means used to achieve the required signals. The $Y$ matrix operates in the manner previously discussed. The $Y$ inverter stage is usually a simple inverting amplifier that changes the polarity of the signal. This inverted signal is then matrixed with the output of the red and blue channels to obtain the $R$-$Y$ and $B$-$Y$ signals.

The $Y$ signal is, of course, transmitted unaltered, but the $R$-$Y$ and $B$-$Y$ must be so encoded that they are transmitted as phase modulation of the 3.579545-megahertz subcarrier.

To facilitate transmitting both color signals as a function of the subcarrier frequency, the incoming subcarrier is divided into two separate signals, both identical in amplitude and phase (Figure 10-19). One of these signals is allowed to proceed unaltered to the *in-phase* modulator, while the other passes through a 90-degree phase shifter before being applied to the *quadrature* modulator. Both modulators are usually identical. The difference in terminology simply reflects the phase difference between the two subcarrier inputs. The $R$-$Y$ and $B$-$Y$ signals are used to modulate the subcarriers. However, the modulator output is sideband modulated only. The carrier is almost completely suppressed.

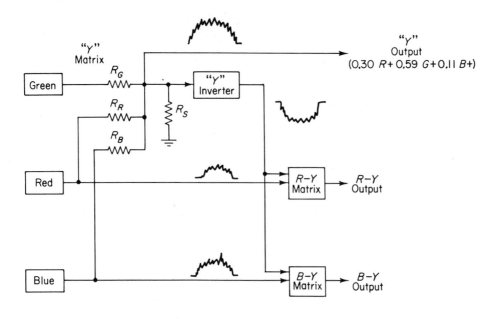

Inverting "Y" and Adding it to the Red and
Blue Signals Removes Brightness Information
from the Red and Blue Signals. The R−Y and B−Y
Outputs Represent Color Information Only.

FIGURE 10-18. Generation of R-Y and B-Y signals.

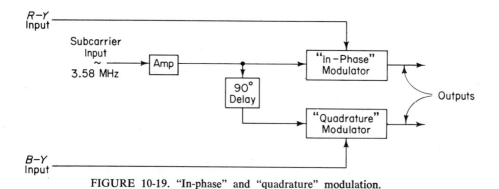

FIGURE 10-19. "In-phase" and "quadrature" modulation.

Figure 10-20 shows a vector representation of the two subcarrier
signals after the delay circuitry has separated them by an angular difference
of 90 degrees. The resultant vector R is a produce of the individual amplitudes of the R-Y and B-Y vectors. The length of R denotes amplitude of

the resultant signal, and its angular displacement represents the phase of the signal. Thus, by manipulating the amplitude of the *R-Y* and *B-Y* vectors, which are displaced from each other by a fixed 90 degrees, it is possible to produce a resultant vector that is variable in amplitude and adjustable in phase throughout a 90-degree arc. For instance, if the *R-Y* vector were reduced to half its amplitude, as shown in dotted lines, the amplitude of the resultant, *R*, would decrease in amplitude and change its phase. This vector representation illustrates exactly what happens when two sinewaves with a 90-degree relative displacement are summed together. If the two signals are of the same amplitude, the resultant signal will be larger in amplitude and possess a phase characteristic that lies directly between that of the two original signals. This is illustrated in Figure 10-21.

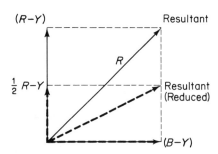

FIGURE 10-20. Changing the amplitude of the *R-Y* vector changes the angle and amplitude of the resultant vector.

Using Figure 10-21 to illustrate the 90-degree phase-shift capability of the resultant sinewave, it is only necessary to visualize wave *A* being reduced in amplitude until it maintains a constant zero potential. Obviously, under these conditions the resultant waveform will be identical to waveform *B*.

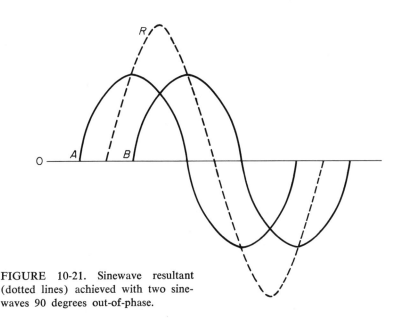

FIGURE 10-21. Sinewave resultant (dotted lines) achieved with two sinewaves 90 degrees out-of-phase.

Thus, there has been a 45-degree phase shift in the resultant waveform, and its amplitude has been reduced to that of waveform *B*.

If waveform *A* were restored to its original amplitude and waveform *B* were reduced to zero, the resultant waveform in Figure 10-21 would alter its phase in the opposite direction until it matched that of waveform *A*. Thus, it can be seen that the vector representation of Figure 10-20 is valid and truly representative of the resultant waveform configuration.

Figure 10-22 illustrates an expanded vector diagram which incorporates the added dimensions of a $-(R\text{-}Y)$ and $-(B\text{-}Y)$. With these additional possibilities, it is possible to generate a resultant vector that has a 360-degree capability. Driving the modulators of Figure 10-19 with positive and negative-going voltages will give this capability.

All of this seems logical enough, but just how does it relate to the color signal? By way of example, assume that a red scene is being scanned by the color camera. From what was discussed earlier it is understood that the two resultant *R-Y* and *B-Y* signals will be:

$$R\text{-}Y = 1 - (0.30R + 0.59G + 0.11B) = 0.70$$

since *G* and *B* equal zero when only red is being scanned.

$$B\text{-}Y = 0 - (0.30R + 0.59G + 0.11B) = 0.30$$

since *G* and *R* equal zero when only blue is being scanned. Thus, plotting a $-0.70$ on the *R-Y* axis and a $-0.30$ on the *B-Y* axis, a resultant vector is achieved that represents the color red.

It is important to note here that the above formulas must be altered by introducing a modifying factor into the *R-Y* and *B-Y* equations. The *R-Y* answer is always multiplied by a factor of 0.877 and the *B-Y* is always multiplied by a factor of 0.493. The reasons for this are entirely practical. If it is not done, the resultant vectors vary considerably in length from one color to another even though they represent thoroughly saturated colors. The modifying process maintains more nearly consistent amplitudes of the chrominance signal for saturated signals, thus alleviating possible problems in modulation. It is a fairly simple matter to restore the vectors to their original configuration in a receiver by reversing this process.

Using the established modifying multipliers, the answer to the previous problem becomes:

$$R\text{-}Y = 0.70 \times 0.877 = 0.614$$
$$B\text{-}Y = 0.30 \times 0.493 = 0.148$$

Thus, the resultant vector appears as in Figure 10-22. The phase angle that represents red is established at a 14-degree angle from the $R$-$Y$ axis.

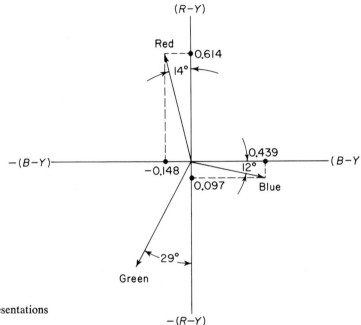

FIGURE 10-22. Vector representations for red, blue, and green.

To compute the angular location of blue, assume that the color camera is scanning an image that contains only blue. Using the established formulas, the vector is plotted as follows:

$$R\text{-}Y = 0 - (0.30R + 0.59G + 0.11B) = -0.11$$
$$B\text{-}Y = 1 - (0.30R + 0.59G + 0.11B) = 0.89$$

since $R$ and $G$ equal zero when only blue is being scanned.
Modified, these become:

$$R\text{-}Y = -0.11 \times 0.877 = 0.096$$
$$B\text{-}Y = \phantom{-}0.89 \times 0.493 = 0.439$$

The resultant vector is shown in Figure 10-22.

The amplitude of the vectors reflects the saturation of the color. The vectors shown assume fully saturated colors that give a 1-volt output amplitude from the camera.

All of the colors that are generated by the color camera may be expressed in this manner. The angular displacement of those that are normally used as reference are shown in Figure 10-23.

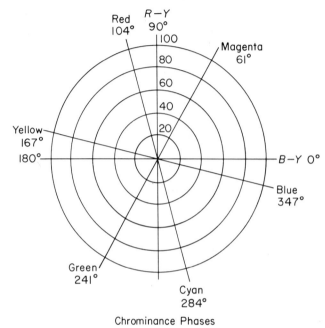

Chrominance Phases

FIGURE 10-23. Vector representation of primary colors and their complements.

## The *I* and *Q* Signals

It has been previously mentioned that the human eye tends to lose color recognition as object size is reduced. Objects below a certain finite size are interpreted by the eye as luminance information only and are seen as different shades of gray. For this reason, in color television transmission it is unnecessary to transmit any color information that lies above 1.5 megahertz. Thus, only those scene elements that produce a video frequency below 1.5 megahertz are transmitted as color. All resolution elements that constitute higher frequencies are transmitted as a luminance signal only.

Closely associated with the above characteristics of the human eye is the fact that relatively small elements of the color picture that are represented by frequencies of approximately 0.5 to 1.5 megahertz may be reproduced to the satisfaction of the eye by utilizing only two primary colors. It has been determined that, if two artificial primaries that are red-orange and blue-green are used for these small areas, the color rendition of the final picture is quite acceptable. This would seem to indicate that small areas of color could be represented along a single axis of the vector diagram if that axis passed through the red-orange and blue-green vector co-ordinates.

Figure 10-24 illustrates a system wherein two new co-ordinate axes have been created—the *I* and *Q* axes. They are located 90 degrees apart as were the *R-Y* and *B-Y* axes, but have been rotated 57 degrees with respect to the *R-Y* and *B-Y* axes. In practice this can be done by simply delaying the two original subcarrier waveforms by a period of time which corresponds to 57 degrees. Figure 10-24 illustrates this in simplified form.

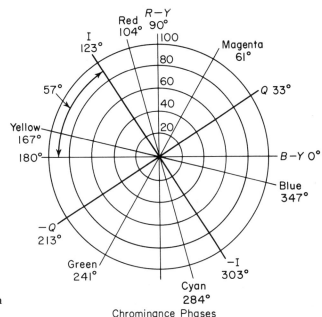

FIGURE 10-24. *I* and *Q* axes establish different references.

Chrominance Phases

Now the *I* axis passes through the red-orange and blue-green areas, and this allows its use as a single co-ordinate axis for all color information that is generated in the 0.5- to 1.5-megahertz frequencies. The *Q* informa-tion is bandwidth limited at 0.5 megahertz.

To restate the above—a three-primary system of color generation is needed for all video frequencies up to 0.5 megahertz, and this is achieved by using two co-ordinate axes (or fixed-phase subcarriers) to generate the correct resultant vector (or resultant phase). The phase of the resultant determines color, or hue, and its amplitude relates saturation. For those frequencies which lie between 0.5 and 1.5 megahertz, it is not necessary to employ two co-ordinate axes because of the limitations of the human eye, and only the *I* axis is used as a reference. Thus, picture elements which lie in this frequency range are transmitted as red-orange or blue-green colors only. Finally, all video information that is above 1.5 mega-hertz is transmitted solely as luminance information, and these frequencies extend to the 4.2-megahertz bandwidth limit.

Figure 10-25 illustrates a bandwidth diagram of the color television signal, showing the sidebands generated by the color information. The *Q* sidebands extend approximately 0.5 megahertz to both sides of the sub-carrier. The *I* sidebands extend 1.5 megahertz below the subcarrier fre-quency, but are suppressed at about 0.5 megahertz above the subcarrier.

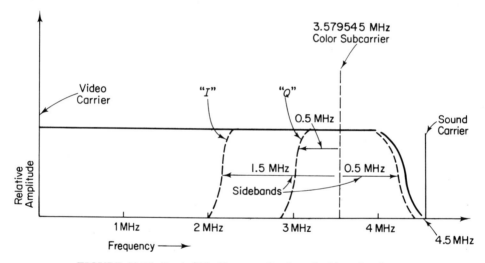

FIGURE 10-25. Bandwidth diagram of color television signal.

Because of the versatility of matrix networks, it is possible to directly generate voltage waveforms which, when applied to the *in-phase* and *quadrature* modulators, will generate the *I* and *Q* signals without the neces-sity of first deriving the *R-Y* and *B-Y* signals. Figure 10-26 shows such a method. The matrix values are such that the *I* and *Q* waveforms that result take the form:

$$I = 0.60R - 0.27G - 0.32B$$
$$Q = 0.21R - 0.52G + 0.31B$$

(These two equations can also be derived by adding approximately 41 per cent of the *B-Y* signal to 48 per cent of the *R-Y* to get the *Q* signal, and adding 27 per cent of the *B-Y* to 74 per cent of the *R-Y* to get the *I* signal.)

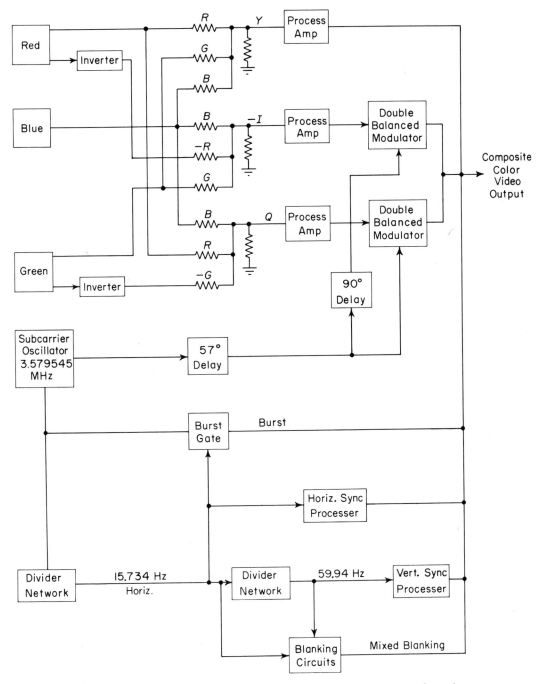

FIGURE 10-26. Block diagram illustrates generation of a composite color television signal.

If the *I* and *Q* waveforms thus developed are added to the luminance signal, the result is a composite signal that contains the brightness, hue, and saturation information that was present in the original scene.

## THE COLOR-BURST AND SYNC SIGNALS

It must be remembered that the subcarrier is not actually transmitted. Only the sidebands that contain the color information are utilized. Subcarrier suppression reduces the energy content of the chrominance information considerably, and this in turn reduces the visual disturbance of the chrominance signal on monochrome receivers. Because of the lack of a subcarrier to use as reference for color demodulation, it is necessary to generate a subcarrier frequency in the receiver and use it for demodulation purposes. It is of the utmost importance that this signal be exactly the same frequency and phase as that of the suppressed subcarrier. To assure correct frequency and phase, a sample of the subcarrier is sent to the receiver by inserting a few cycles of the 3.579545-megahertz signal from the camera system onto the composite color video waveform. Eight or nine cycles are gated onto the video just after the horizontal-sync pulse, as shown in Figure 10-27. This short sample of the subcarrier is known as the *burst* signal. Because it interrupts the back porch portion of the horizontal blanking, the area between the end of horizontal sync and the beginning of burst is called the *breezeway*.

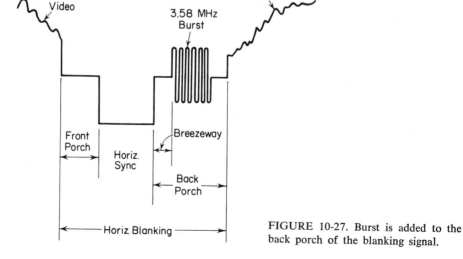

FIGURE 10-27. Burst is added to the back porch of the blanking signal.

The burst signal is used in the color receiver to control the phase of the receiver subcarrier oscillator. Generally, incoming burst signal is compared to the oscillator output, and a dc "difference" signal is generated when an out-of-phase condition exists. This voltage is then applied to the oscillator as bias and serves to pull it back into the proper phase relationship. Figure 10-28 illustrates the phase relationship of the burst signal with that of the *I* and *Q* signals. (Also see Figure 10-26.)

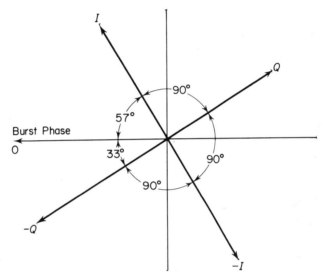

FIGURE 10-28. Burst phase referenced to *I* and *Q* signals.

Because the color burst signal must maintain a constant phase relationship with the scanning signals to assure proper frequency interleaving, the horizontal and vertical synchronizing pulses are also derived with reference to the subcarrier frequency. Countdown circuitry is generally incorporated to divide the 3.579545 megahertz down to the 15,734-hertz horizontal frequency. With such a rigid relationship established, the phase characteristics of the burst and sync signals remain constant with respect to one another.

The vertical sync is, of course, derived from the horizontal frequency.

# THE COLOR CAMERA AND
# ASSOCIATED CIRCUITS

The color television camera is a much more complex device than its monochrome counterpart. Essentially, it consists of multiple monochrome cameras that have been carefully matched and adjusted to provide outputs which differ only in amplitude as a function of the color content of the scene being televised. In this chapter we shall not be concerned with the matrixing or modulation of the color signal to obtain the NTSC composite video signal, but will consider only the signals as developed by the individual color and luminance channels within the camera system.

It has been previously demonstrated that the optical arrangement in a color camera assures that each camera tube receives scene information that represents only one of the three basic primary colors that are used and, in a three-tube camera, the three outputs are matrixed together to provide a luminance signal to convey the brightness information.

Many color cameras (Figure 11-1) utilize a fourth camera tube to generate the luminance signal. This method has certain advantages over the three-tube system in some areas and disadvantages in other areas.

The four-tube color camera has a disadvantage in that it must split the incoming light into four paths instead of three, thus lowering the amount of light that any one tube receives. This could have the effect of lowering the signal-to-noise ratio of the camera. On the other hand, the luminance signal in the four-tube camera is not dependent upon a matrixed combination of three other signals which may not be entirely coincident because of nonlinearities and misregistration within the system. This offers better resolution in the luminance signal.

FIGURE 11-1. RCA TK-42 color camera utilizes three vidicons and an image-orthicon.

Many four-tube cameras use vidicons or similar devices for the three color channels, and an image-orthicon or larger vidicon for the luminance channel. The image-orthicon, of course, provides good low-light sensitivity and its "overshoot" characteristic on abrupt edges within the image gives an apparent sharpness and clarity to the final picture. To achieve this same effect in a three-tube camera system requires the use of an electronic means of signal enhancement. Thus, the extra cost and complexity of a four-tube color camera finds an approximate equal in a three-tube system that uses electronic enhancement to achieve sharpness of image. There have been, and will probably continue to be, many heated discussions concerning the relative merits of both systems.

## OPTICS

Figure 11-2 illustrates an arrangement whereby three vidicons are used for the color channels and an image-orthicon generates the luminance signal. In addition to mirrors, prisms are used to alter the directional characteristics of the light path.

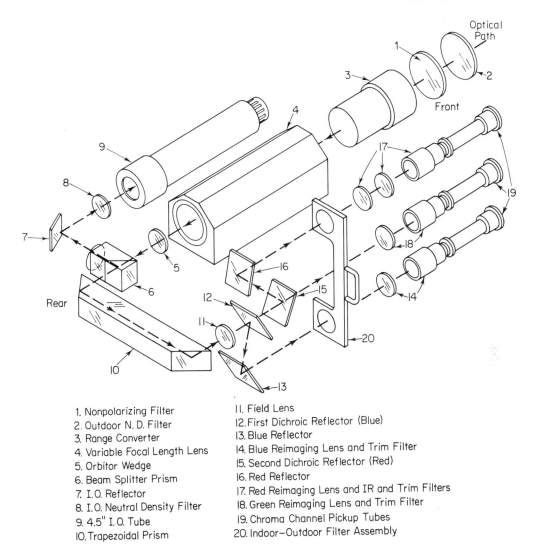

COURTESY RCA.

FIGURE 11-2. Optical arrangement in RCA TK-42 color camera.

1. Nonpolarizing Filter
2. Outdoor N. D. Filter
3. Range Converter
4. Variable Focal Length Lens
5. Orbitor Wedge
6. Beam Splitter Prism
7. I.O. Reflector
8. I.O. Neutral Density Filter
9. 4.5" I. O. Tube
10. Trapezoidal Prism

11. Field Lens
12. First Dichroic Reflector (Blue)
13. Blue Reflector
14. Blue Reimaging Lens and Trim Filter
15. Second Dichroic Reflector (Red)
16. Red Reflector
17. Red Reimaging Lens and IR and Trim Filters
18. Green Reimaging Lens and Trim Filter
19. Chroma Channel Pickup Tubes
20. Indoor–Outdoor Filter Assembly

The optical diagram represents the system used in the RCA TK-42 camera and may be considered as a good example of the complexity of such systems. Of course, configuration varies considerably with different manufacturers and the use for which the camera is intended.

Referring to the diagram, some of the main components of the system are as follows:

VARIABLE FOCAL-LENGTH LENS. This is more commonly called a zoom lens, and is described in Chapter Three.

NONPOLARIZING FILTER. The nonpolarizing filter is used to free the luminance channel from polarizing effects. Its main purpose is to keep the general luminance information from varying when the camera is used in situations of varying polarized light.

OUTDOOR NEUTRAL-DENSITY FILTERS. This filter is inserted into the main optical path when the camera is operated in areas of high light level. It simply attenuates the amount of light that enters the system, but does not alter the color content.

RANGE CONVERTER. The range converter operates in conjunction with the zoom lens to alter the focal length of the lens system. For instance, if the zoom lens has the capability of a 5:1 ratio, 4 to 20 millimeters, inserting the range converter will change this to 8 to 40 millimeters.

ORBITER WEDGE. The orbiter wedge is used on some cameras to prevent image burn-in of the pickup tubes photocathodes when the camera views the same object for a period of time. If the camera output is not being viewed at any particular time, the orbiter, which is a motor driven optical wedge, may be energized to rotate at about 1 rpm. The rotation gives an approximate 5 per cent circular motion of the picture, preventing burn-in. When the camera is selected for use, the orbiter is, of course, deactivated.

BEAM SPLITTER PRISM. The optical path continues through a beam splitter prism where light for the luminance channel is reflected to the image-orthicon tube. The prism also includes a filter to assure the correct amount of light for the image-orthicon.

IMAGE-ORTHICON REFLECTOR. This front-surface mirror directs the luminance image to the image-orthicon tube.

TRAPEZOIDAL PRISM. The trapezoidal prism causes the incoming light to be reversed in direction and offset to one side. It also lengthens the back focal distance and allows the placement of a field lens.

FIELD LENS. The field lens provides for a concise arrangement of optical components in the three chroma channels.

RED AND BLUE DICHROIC REFLECTORS. These dichroic mirrors serve to reflect light of a particular color and allow all other light to pass through, thus essentially splitting the light path into three separate elements by color. Red and blue are separated by reflection and the green that is left passes through to its own reimaging lens and camera tube.

REIMAGING LENSES. These lenses reduce the image to the proper size for use on the chroma pickup tubes.

INDOOR/OUTDOOR FILTER. This filter allows for a quick change between indoor and outdoor operation. The filter compensates for the difference in color temperature that exists between natural light and artificial light.

The optical system illustrated is from a camera designed for use in commercial broadcasting applications. Three-tube cameras and color cameras designed for closed circuit use typically employ optical systems which are less complex.

It is imperative that the optical alignment of the light paths be precise. If they are not, the reproduced color image will be *misregistered*. That is, each image will not be located at exactly the same point on the camera tubes and when they are combined to form the final color image, a blurred presentation will result.

## THE SCANNING SYSTEM

Scanning, as performed in a color camera, is generally accomplished in much the same manner as it is in monochrome systems. Since there are multiple camera tubes and yokes in the color camera, it is imperative that all scanning waveforms be precisely synchronous. For this reason, the respective deflection signals are generally derived from a single oscillator.

It is also necessary to use deflection yokes that have been matched for performance so that variations in raster geometry do not cause serious misregistration problems. Each deflection yoke is usually provided with its own size and linearity controls so that proper adjustment of these parameters, together with centering capability, will allow the individual scanning systems to be critically aligned.

Figure 11-3 illustrates a system for generating horizontal scanning waveforms for a three-tube color camera. The horizontal drive from a sync generator is applied to the base of $Q1$ which amplifies the pulse to use as drive for $Q2$. $Q2$ is normally biased *on,* and the applied pulse causes it to turn *off,* thus allowing the energy stored in all three yokes and the saturable reactor, $L2$, to generate a flyback pulse. Circuit operation is very similar to that discussed in Chapter Five. $L2$ functions as the master control for horizontal linearity and it affects the three parallel yokes in an identical manner. The green channel linearity is set using $L2$, and no other adjustments are provided. However, the red and blue channels contain additional linearity controls which allow each to be adjusted to match the linearity of the green channel.

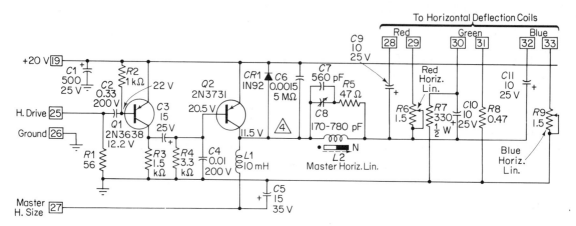

COURTESY COHU ELECTRONICS, INC.

FIGURE 11-3. Horizontal deflection circuit drives three vidicon deflection coils.

The master horizontal size and centering controls are used to adjust the green channel for proper dimension and placement, and additional controls at the yoke (not shown) serve as trimmer adjustments for the red and blue channels.

A method for generating vertical deflection is shown in Figure 11-4. An emitter follower, $Q8$, isolates the vertical drive input and provides a low-impedance source to trigger the unijunction oscillator, $Q9$. The sawtooth waveform generated at the emitter of $Q9$ is linearized and provided with additional current gain through the circuitry of $Q10$ and is applied to the vertical driver, $Q11$, through the master vertical-size potentiometer, $R37$. $Q12$ serves as the vertical output stage and drives all three color yokes. Because green is used as the master channel in this case, the master vertical size is used for the green size-control, and the red and blue yokes

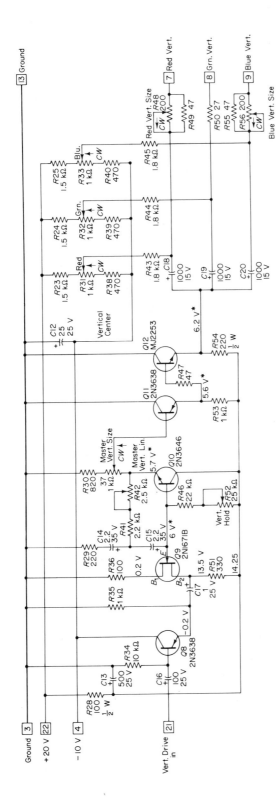

FIGURE 11-4. Vertical deflection circuit for a three-vidicon color camera.

are trimmed for conformance with series potentiometers $R48$ and $R56$. Vertical centering for each channel is achieved by the three centering potentiometers which provide the proper dc current for the red, green, and blue yokes. Because the vertical yoke is essentially a resistive load at the low vertical frequency, and because of the matched characteristics of the three yokes, no individual linearity controls are required.

Figure 11-5 shows a pattern that is commonly used for registration of a color camera. The many horizontal and vertical lines give good reference points across the raster with which to check for any misregistration. A common procedure would be to display the green channel and one of the others, while switching out the remaining channel (or channels, in the case of a four-tube camera). With the two channels superimposed, size, centering, and linearity controls can then be adjusted to bring the two images into coincidence over the entire raster. This procedure would be repeated for the other channel(s) until all were properly registered.

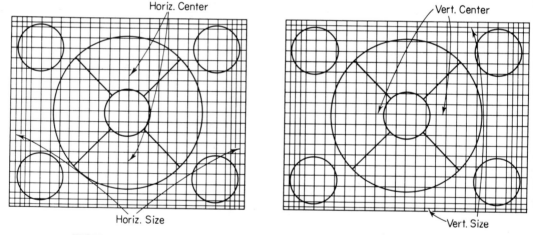

FIGURE 11-5. Pattern used for registration of a color camera.

## SHADING GENERATORS

Due to the many irregularities that may occur within the camera pickup tubes and their associated deflection systems, the resultant video presentation may appear shaded in certain areas of the raster. As an example, the output of a particular camera tube may cause the upper left-hand corner of a presentation to be somewhat darker than the rest of the picture. In another case, the entire right-hand side of the picture may appear brighter than the rest of the image even when a flat background is being televised.

While shading may certainly pose problems in monochrome television systems, its effects in a color camera are completely unacceptable. Since the color camera relies on an accurate percentage of each color channel to compose a complete color picture, obviously any variations in these signals caused by shading will upset the desired ratio between signals. To illustrate, if the red channel produces a corner that is brighter than it normally should be for a particular scene, the ultimate display will show an unnatural predominance of red in that corner. Thus, if the red, green, and blue channels were called upon to reproduce a white scene, the unbalance caused by the corner shading in the red channel would produce a red hue in the corner.

To correct for shading irregularities it is a common practice to employ *shading generators* that supply waveforms to the cathodes of the camera tubes to offset the variations that are present because of shading irregularities. Such waveforms modulate the beam, increasing its intensity to compensate for a dark area of the raster and decreasing the intensity in areas which are too bright.

Figure 11-6 is a block diagram of circuitry that will generate the proper signals for shading compensation. It develops sawtooth and parabolic waveforms at the horizontal and vertical rate. These waveforms, when mixed in the proper proportions, can compensate for shading problems of varying degree and configuration. The horizontal trapezoid generator develops a sawtooth waveform during active scanning time and provides both positive- and negative-going outputs from the phase splitter. Thus, horizontal sawtooth waveforms of opposite polarity are applied across the *horizontal saw controls*. A horizontal parabola waveform, occurring during the horizontal scanning time, is generated by the integrator $(Q1/Q2)$ and applied across the *horizontal parabola* controls. The action of the vertical circuits is similar.

Using the circuit shown, shading correction can be applied to any of the four tubes (assuming a four-tube camera) in the form of a composite waveform that may embody any or all of the available waveforms.

Figure 11-7 illustrates circuitry used to accomplish the generation of a horizontal parabolic waveform. A sawtooth signal is fed into the feedback pair of parabolic generators $Q1$ and $Q2$ and is integrated due to the capacity of $C3$ which lies in the feedback path. Since the integral of a

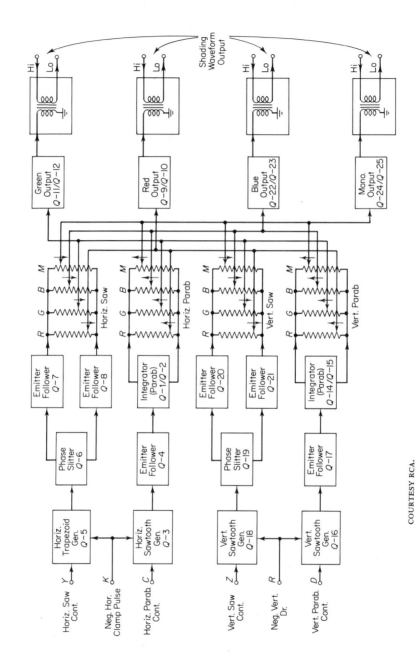

COURTESY RCA.

FIGURE 11-6. Block diagram of a shading generator.

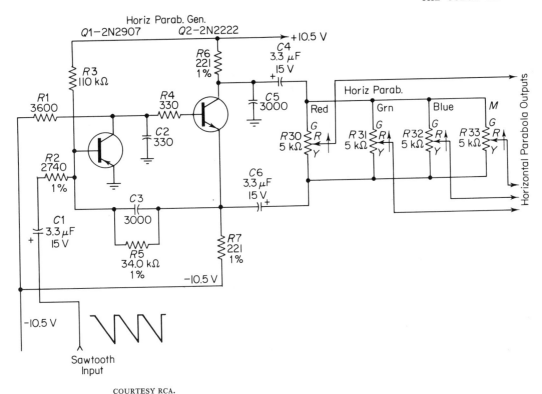

COURTESY RCA.

FIGURE 11-7. Horizontal parabola generation for a four-tube camera.

sawtooth is a parabolic waveform, it is this signal that is applied to the output transistor. Because the collector and emitter loads of $Q2$ are equal, the signals developed at each terminal are equal in amplitude and opposite in polarity. Applying these signals across the horizontal parabola potentiometers gives a continuously adjustable waveform that may be routed to each of the camera tubes as necessary.

Vertical parabolic generators operate in an identical manner with the exception of some larger time constants to accommodate the lower frequency.

A typical example of a sawtooth generator is illustrated in Figure 11-8. The time constants shown are those needed for a sawtooth waveform occurring at the vertical rate. Transistor $Q18$ is normally cut off, and $C32$ charges at a linear rate determined by the resistance value of $R75$ and the potential applied to it. When an input pulse, occurring at the vertical rate, turns $Q18$ on momentarily, capacitor $C32$ discharges. This action creates

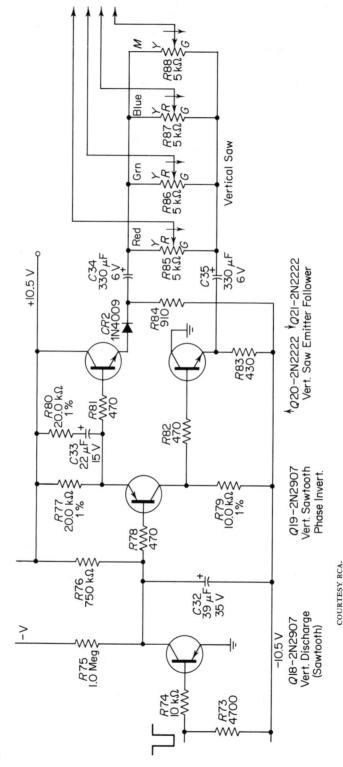

COURTESY RCA.

FIGURE 11-8. Sawtooth generator for shading generator.

272

a sawtooth waveform. When the input pulse is removed, $C32$ again begins charging and the cycle repeats itself.

The output of $Q18$ is fed into a paraphase amplifier, $Q19$, to provide two signals that are equal but opposite in polarity. These two signals are then fed through emitter followers $Q20$ and $Q21$ to the vertical sawtooth controls.

When it is desired to make shading adjustments on a color camera system, it is necessary to aim the camera at a light source that exhibits minimal shading characteristics itself. The camera lens should be set to a high f/stop to minimize any vignetting or other aberrations, and the focus of the lens should be adjusted so that no sharp images are apparent. All of these factors serve to assure that the camera tubes will be flooded with a diffuse, flat light.

With an oscilloscope set to view the video outputs at the horizontal rate, one of the video waveforms may be selected and the horizontal sawtooth and parabolic controls should then be adjusted to provide a flat line presentation, eliminating any amplitude responses that are caused by shading. When each channel has been adjusted at the horizontal rate, the waveform monitor should then be set to view the video waveforms at the vertical rate. The vertical sawtooth and parabolic controls may then be adjusted to provide for a flat vertical waveform presentation from each channel.

Once the above adjustments have been made, a final alignment is generally made by superimposing the video waveforms one upon the other and varying the shading controls to obtain a critical match between the waveforms of each channel.

## GAMMA CORRECTION

The inclusion of gamma correction circuits in color cameras is especially important. In three-tube color cameras they are needed to compensate for the differences in light transfer characteristics that will exist between tubes of the same type. Although tube manufacturers try to hold these tube parameters within close tolerances, it is very rare indeed to find two tubes that appear to have identical characteristics. Because of the necessity of having three identical channels to process the color scene into usable video signals, gamma correction is used to adjust the signal characteristics in a manner that will make it appear that all three camera tubes have the same light-transfer characteristics.

Gamma correction also performs these same functions in a four-tube camera, but finds added significance where a different type tube is used for the luminance channel. For example, consider cameras that use three vidicons and an image-orthicon as the tube complement. The image-orthicon has a light transfer characteristic that differs considerably from that of the vidicon. Therefore, it becomes the task of gamma circuitry to alter the signal electronically until the luminance signal is composed of amplitude variations essentially the same as would be achieved if the three color channels were matrixed to obtain luminance. This is a formidable task, and is never completely achieved in practice.

To say that gamma correction circuitry corrects for the differences in transfer characteristics of pickup tubes is, of course, not enough. The gamma correction circuitry must also correct for the nonlinearity of light output from the color picture tube. As with monochrome television, it is unfortunately true that an increase in the video signal that drives the picture tube does not result in a completely proportional increase in light output. Another source of luminance distortion may occur in the video processing amplifiers themselves, where high-level amplification may cause either the white or black portions of the signal to be compressed.

It is a standard practice to compensate for all of the aforementioned distortions by the use of gamma correction circuits in the camera system. Since the characteristics of the receiver picture tube are known, correction may be introduced which will assure that the final, overall light transfer characteristic of the complete television system, from lens to picture tube, is relatively linear.

Some typical gamma correction circuits were shown in Chapter Six.

## ELECTRONIC SIGNAL ENHANCEMENT

In Chapter Six, aperture correction was discussed as a means of enhancing the video presentation by electronically accenting either the high-frequency elements or abrupt amplitude transitions within the video signal. The result of such means was a "sharper" or "crisper" picture. In those illustrations, however, the enhancing effect was achieved on horizontal resolution elements only.

In color television cameras of the three-tube variety, it is often necessary to introduce signal enhancement that affects both horizontal and vertical picture detail. If this is not done, the three-tube camera will generally produce a picture that appears softer, or less detailed, than a four-tube camera—especially one that uses an image-orthicon as the luminance source.

As a rather broad generalization, it can be said that electronic image enhancement is used to simulate one of the desirable characteristics of the image-orthicon; i.e., the signal overshoot that occurs as the beam moves from one area of contrast to another of different relative content.

At present, the most widely used means of image enhancement is accomplished by delaying the video and comparing it with signal elements which have also been derived by delay techniques. For instance, correction waveforms for vertical picture elements are derived by comparing a single horizontal line of video with the line that directly preceded it and that which directly follows it. Any differences in content are detected and a correction (enhancement) signal is generated. When this is later added to the video, a very definite sharpening of the picture is observed.

Figure 11-9 is a block diagram of the CBS Laboratories' Image Enhancer. The video applied at the input is amplified and fed into a modulator which double-sideband amplitude-modulates it on a 30-megahertz carrier frequency being generated by the crystal oscillator. This modulation allows the video to be processed by various delay networks without altering the content of the video signal. Upon being delayed, the video is detected and restored to its original form. It may be seen that the *top* video is not delayed, the *main* video is delayed by one horizontal line (63 microseconds), and the *bottom* video is delayed by two horizontal lines (126 microseconds). The demodulated *top* video and *bottom* video are each routed to separate differential amplifiers through gain controls that are used to equalize their amplitudes. The *main* video is also applied to each of the differential amplifiers, with the result that the top differential amplifier subtracts the *top* video from the *main* video and the resulting output is $M - T$.

The lower differential amplifier subtracts the *bottom* video from the *main* video and results in an output that is $M - B$.

When the two outputs are added together at the output of the differential amplifiers, the final result may be expressed as $2M - T - B$. A vertical equalizing gain control provided at this point effectively varies the amount of vertical detail information that will be available.

From the above, it may be apparent that the video information reaching the differential amplifiers consists of three separate, sequential horizontal lines of video. If there is no difference between the signals, simple

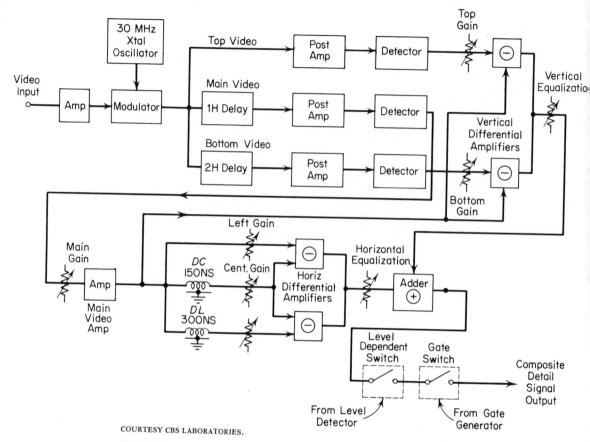

FIGURE 11-9. Simplified block diagram of CBS Laboratories' image enhancer.

mathematics shows that $2M - T - B$ will equal a zero output. However, if $T$ or $B$ varies with respect to $M$, an output will be effected and this will then be used to generate the correction signal that is used for vertical aperture correction.

If attention is returned to the *main* video amplifier, it can be seen that the *main* video signal is also applied to the horizontal aperture equalizer. Here the *main* video is separated into three paths, using the same procedure that was shown in the vertical circuits. In this case, there is an undelayed signal (called the *left* video signal), a signal delayed by 150 nanoseconds (called the *center* video signal), and a signal delayed by 300 nanoseconds (called the *right* video signal). The *center* and *left* video output signals are subtracted in a difference amplifier whose output is then $C - L$. The *center* and *right* video signals are similarly processed and the output of the second difference amplifier is $C - R$.

When the two outputs from the horizontal differential amplifiers are added together, a resultant signal of $2C - L - R$ is achieved. Thus, when there is no variation in signal level during the course of a horizontal line of video, *C, L,* and *R* are equal in amplitude and the resultant output is zero. However, when an amplitude variation occurs there is, for a period of 150 nanoseconds, a difference in potential between the *left* video signal and the *center* video signal, and a correction signal is developed in the top differential amplifier. Then, for an additional 150 nanoseconds, a difference in potential exists between the *right* video signal and the *center* video signal and an equal, but opposite polarity, signal is developed by the bottom differential amplifier.

The outputs from the vertical and horizontal differential amplifier blocks are summed in the adder block into a composite detail signal output. This signal is passed through two separate gates, the *level dependent* switch and the *gate* switch. The level dependent switch is included to inhibit the *detail* signal during the time when the *main* video signal is darker than any preset shade of gray. This acts to eliminate noise in the darker areas of the picture. The *gate* switch acts to inhibit the *detail* signal during the period from just slightly before horizontal blanking until just slightly after the horizontal blanking interval. It also inhibits the signal in a similar manner during the vertical blanking interval. This eliminates the possibility of transient generation at the beginning and end of normal horizontal and vertical blanking intervals.

Figure 11-10 is a schematic representation of the double-sideband modulator which was shown in the block diagram of Figure 11-9. The video input, a 0.5- to 2.0-volt signal, is amplified by $Q1$, $Q2$, $Q3$, and $Q4$ to an approximate 4-volt signal at $TP2$. The potentiometer $R3$ acts as a gain control while $R7$ is adjusted for best amplifier linearity.

Transistor $Q5$ forms the crystal-controlled oscillator that generates the 30-megahertz carrier frequency. It couples signals via transformer $T2$ to the base elements of transistors $Q6$ and $Q7$. Transistors $Q6$ and $Q7$ operate in a push-pull arrangement that chops the video at a 30-megahertz rate. $Q8$ and $Q9$ act as RF output amplifiers for the 30-megahertz carrier frequency, and since the video from $Q4$ is applied to both transistors in phase, it is seen as a common mode signal across the output transformer $T1$ and is cancelled. $T1$, in conjunction with $C14$ and $C17$, acts as a double-tuned

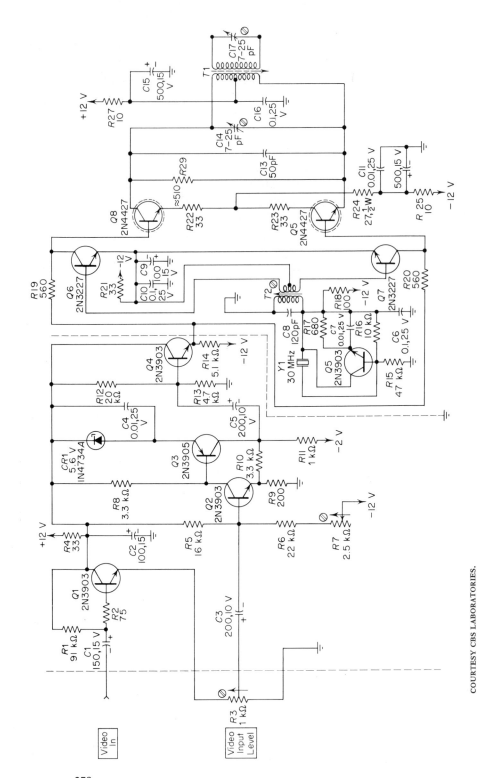

COURTESY CBS LABORATORIES.

FIGURE 11-10. A double-sideband modulator readies signal for delay lines.

bandpass filter, and the output is essentially double-sideband amplitude modulation.

The RF output is applied to the delay lines shown in Figure 11-9. The delayed *bottom* and *main* video and the undelayed *top* video are then applied to the post amplifiers shown in Figure 11-11.

The post amplifiers in Figure 11-11 consist of $Q1$, $Q4$, and $Q7$. Consider first the *top,* or undelayed, video. After being amplified by $Q7$, the modulated signal is routed to detector stage $Q8$ where the *top* video is demodulated and fed into a multisection low-pass pi filter to remove the 30-megahertz carrier from the video signal. From the filter, the video is applied to $Q9$ which acts as an emitter follower to isolate the filter from the output line.

The operation of the *main* and *bottom* signal channels is essentially the same except for the integrated circuit amplifiers, $IC1$ and $IC2$, which are incorporated in each to compensate for the additional attenuation that was experienced in these channels due to the delay line action. Also, it may be noted that additional delay lines are shown in the base circuits of $Q3$ and $Q6$. These are incorporated as trimmers to assure that the *main, top,* and *bottom* video signals are all exactly coincident for proper comparison in the differential amplifiers.

The differential amplifiers, consisting of integrated circuits $IC1$ and $IC2$, are shown in Figure 11-12. The *top* and *main* signals are applied to $IC2$ to produce the $M - T$ signal, and the *main* and *bottom* signals are applied to $IC1$ to yield the $M - B$ signal. The outputs from the two differential amplifiers are summed at the junction of $R9$ and $R10$ through switches $S1$ and $S2$, which provide a means of observing the vertical top, or vertical bottom, correction only.

Transistor $Q1$ forms a low-impedance output stage whose base circuit comprises a low-pass filter with a cutoff frequency of approximately 2 megahertz. The filter is terminated in $R11$, the *vertical equalizing* gain control that adjusts the output amplitude of the vertical correction signal.

The horizontal correction waveform is generated by the circuitry shown in Figure 11-13. The input video signal is amplitude adjusted by $R1$, the *main* gain control, and applied to $IC1$, an RCA CA3001 amplifier. The output signal at pin 8 is applied to the base of $Q1$, which acts as a low-impedance driving source for the delay lines and individual *left, center,* and *right* gain controls. The *left* signal is taken from the arm of $R5$, giving an undelayed input to the differential amplifier $IC3$. The *center* signal is produced by delaying the video signal 150 nanoseconds through delay line

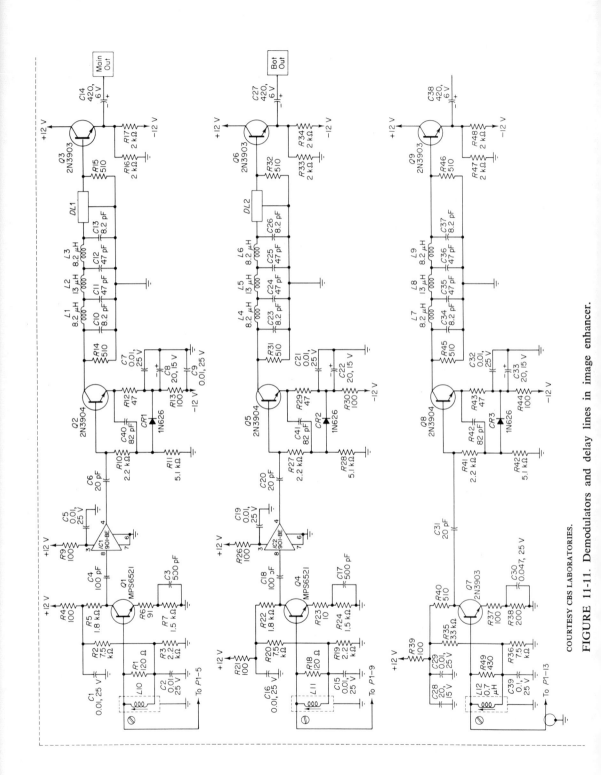

COURTESY CBS LABORATORIES.

FIGURE 11-11. Demodulators and delay lines in image enhancer.

280

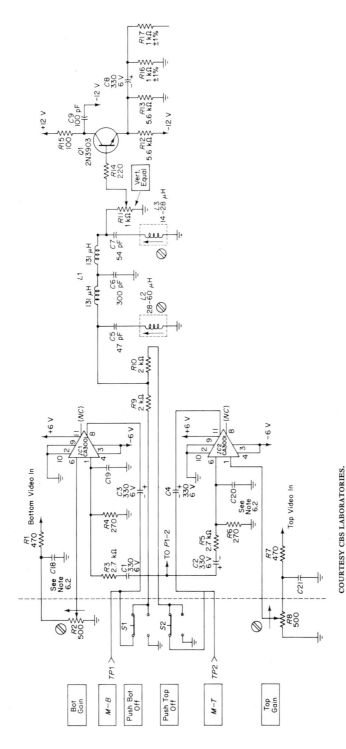

FIGURE 11-12. Differential amplifiers yield difference signals.

281

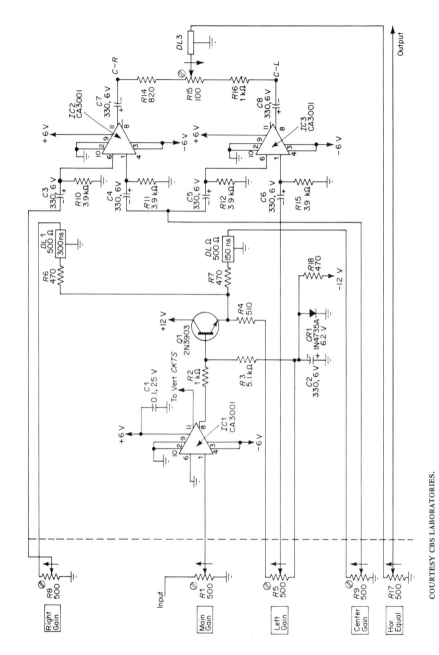

FIGURE 11-13. Circuit provides horizontal enhancement waveforms.

*DL2*, and the *right* video signal is generated by delaying the video signal 300 nanoseconds through delay line *DL1*. The *center* video signal is applied to both differential amplifiers and the *right* video is applied to *IC2* only.

Differential amplifier *IC2* compares the *right* and *center* video signals and provides an output that is the difference between the two, $C - R$. *IC3* similarly provides a $C - L$ signal which is the difference between the *center* and *left* signals. The outputs of the two differential amplifiers are summed to make the *horizontal detail* signal $2C - L - R$.

An additional delay line, *DL3*, is added at the output to make the *horizontal detail* signal exactly coincident with the *vertical detail* signal previously explained. *R17* allows adjustment of the amplitude of the *horizontal detail* signal.

Figure 11-14 schematically illustrates the manner in which the horizontal and vertical detail elements are added. Also shown are the *gate* and *level dependent* switches *Q2* and *Q3*, which serve to inhibit detail output during blanking intervals and when the signal is below specific preset levels. Vertical detail information is applied to integrated-circuit amplifier *IC3* at pin 6 and the horizontal detail information is applied to pin 1. The outputs of the *IC* amplifier at pins 8 and 11 are identical but opposite in polarity. These are applied to two class B amplifiers, *Q5* and *Q6*, which provide the "crispening" characteristics of the output signal.

When the crispening control, *R25*, in Figure 11-14 is positioned so the center arm is fully counterclockwise (arrow indicates clockwise direction of travel), both transistors operate as class B amplifiers and each passes one-half cycle of the incoming video waveform. As *R25* is moved clockwise, the transistors begin to clip the negative excursions of the signal. *Q5* provides an output at its emitter junction that contains only the positive white peaks of the applied detail signal, while *Q6* yields an output at its collector and passes only the positive black peaks of its applied detail signal. *Q6*, therefore, inverts its input signal, which was opposite in polarity to the input of *Q5*, and re-establishes the proper relationship for summing the two outputs at the junction of *R30* and *R31*. When the outputs are added, there results a detail signal with positive white and negative black excursions. Turning the crispening control clockwise causes a slice to be removed from the middle of the detail signal and this process is termed *crispening*.

Gating of the *detail* signal is accomplished by the FET *Q4* during the blanking periods. During these intervals it shorts pins 1 and 6 of the IC

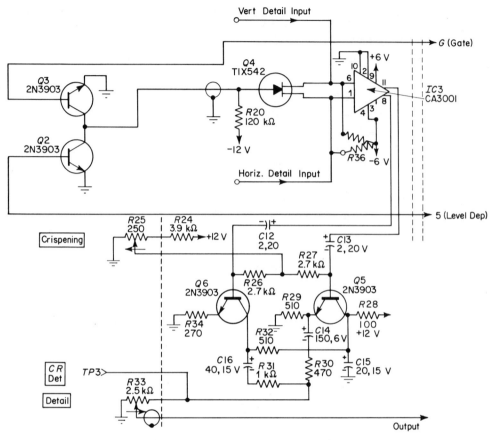

FIGURE 11-14. Horizontal and vertical detail signals are summed by an IC amplifier.

amplifier, and no output from the system is realized. *Q4* is gated by *Q2* and *Q3*, which are controlled in turn by external signals derived from blanking gates and a level dependent amplifier.

It is important to note here that most image enhancing devices used in color television operate by enhancing only the video signal generated in the green channel of the color camera. The signals developed by the red and blue channels are not sampled and consequently no correction signal is generated during signal excursions in these two channels. However, the results of using only the green video signal are quite acceptable, and this is understandable when it is remembered that the green signal comprises 59 per cent of the total luminance signal. The detail signal when added to the luminance signal appears quite natural and creates a sharpness of image otherwise unobtainable in a three-tube color camera.

## ADDITIONAL CIRCUITS

Additional circuitry used in color television cameras may include the synchronizing generators, color modulators or encoders, and sophisticated processing circuits. However, these units are often housed in separate enclosures and may be used with a wide variety of color cameras. For this reason they will be treated separately in the following two chapters.

Circuits that involve video amplifiers and processors, sync and blanking addition, power supplies, etc., are, for the most part, very similar to those used for like functions in monochrome cameras. Attention to the examples given in Chapter Six will give a good indication of the methods employed. Deviations that might occur in color cameras would be influenced only by special requirements, such as bandwidth, critical gamma needs, and the like.

# 12

## SYNC LOCK CIRCUITRY AND
## SUBCARRIER GENERATORS

A sync generator that is to be used with a color television system must conform to much more rigid standards than one designed for monochrome applications. To achieve the necessary frequency interleaving, as described in Chapter Ten, it is necessary that the horizontal and vertical frequencies be an exact submultiple of the color subcarrier frequency. It was previously shown that these are (approximately):

| | |
|---|---|
| Horizontal Frequency | 15.734 hertz |
| Vertical Frequency | 59.94 hertz |

Of course, the horizontal and vertical frequencies must still maintain a constant phase relationship with each other, but equally important in a color system, they must be established in a constant phase relationship with the color subcarrier. Thus, in many systems, a sample of the 3.58-megahertz subcarrier is used to drive a counter chain that counts down to the horizontal frequency and supplies output which will then be used in a phase comparator. The phase comparator will detect any change between the phase characteristics of the horizontal frequency being supplied by the sync generator and that developed by the color subcarrier countdown. An error voltage is then developed that is applied to the sync generator master oscillator to effect any change that may be necessary to maintain the proper phase relationship.

Because it is common procedure in broadcast applications to use remotely generated programs, from a network source for instance, and then

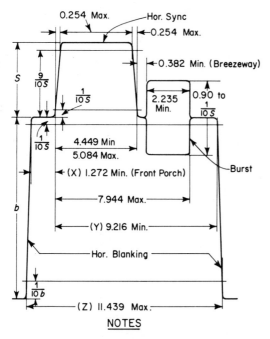

NOTES

1. Subcarrier Freq. = 3,579,545 ± 10 C.P.S.

2. Hor. Freq. = 15,734,264 ± 0.044 C.P.S.

3. Maximum Freq. is Used in Calculating Maximum Durations, Which Are Expressed to The Nearest 1000th Microsecond Below Precise Values

4. Minimum Freq. is Used in Calculating Minimum Durations, Which Are Expressed to The Nearest 1000th Microsecond Above Precise Values

5. $S = 0.4b$ (Nominal)

6. Leading and Trailing Slopes of Hor. Blanking Must Be Steep Enough to Preserve Min. And Max. Values of $(X + Y)$ and $(Z)$ Under All Conditions of Picture Content

FIGURE 12-1. FCC TV sync generator waveform specifications for color transmission.

switch periodically to a locally generated signal (for commercials, station identification, etc.), it is desirable to have a means to lock the horizontal and vertical phase of the locally generated sync to that of the incoming signal. Thus, when switching between the two sources, the transitions may be made with as little disturbance on the receiver screen as possible. When the local and remote program material is in color, it is also necessary to maintain a constant phase relationship between the locally generated sub-carrier and that of the incoming composite color signal.

Equipment that maintains a constant phase relationship, both horizontally and vertically, between two different sync generator systems is called a *sync lock* or *genlock* unit. Subcarrier generators, which generally will also provide the countdown necessary to phase lock a synchronizing generator are called *color standards* or *color subcarrier generators*. Each of these units will normally be designed to work in conjunction with a specific sync generator. To a certain degree, equipment of specific design may not be compatible with that made by a different manufacturer. All will be compatible with standard EIA sync and video waveforms however.

It is the responsibility of the above equipment, together with an appropriate sync generator, to develop a synchronizing waveform like that shown in Figure 12-1, which conforms to the FCC TV sync generator waveform specifications for color transmission.

## SYNC LOCK CIRCUITS

To lock the horizontal and vertical output components of a local sync generator with those that exist on an incoming composite video signal, it is necessary to strip the sync from the video waveform and separate it into its horizontal and vertical components. Figure 12-2 illustrates circuitry that will accomplish the sync separation. Although similar to the method employed in some monitors, some interesting differences may be noted.

The incoming signal is processed by two emitter followers, $Q1$ and $Q3$, to provide the low impedance necessary to drive the horizontal and vertical extractor circuits. In this instance, the incoming waveform has had the video component of the composite signal removed. The vertical-rate pulse is derived by the vertical sync extractor, $CR2$, $CR3$, $Q4$, $Q5$, and $Q6$.

In operation, the normally short-duration negative pulses which constitute the horizontal sync and equalizing pulses build up a slight charge on the positive plate of capacitor $C3$, but it is quickly drained off by the $RC$ time constant determined by the capacitor and $R11$. This negative charge is not sufficient to overcome the reverse bias applied to diode $CR3$ by the application of a negative voltage at its anode through voltage divider $R14$ and $R13$. However, when the vertical sync interval comes along with its wide negative pulses, the negative potential developed at the cathode of $CR3$, together with the action of $L1$ and $C4$, causes a trigger to be

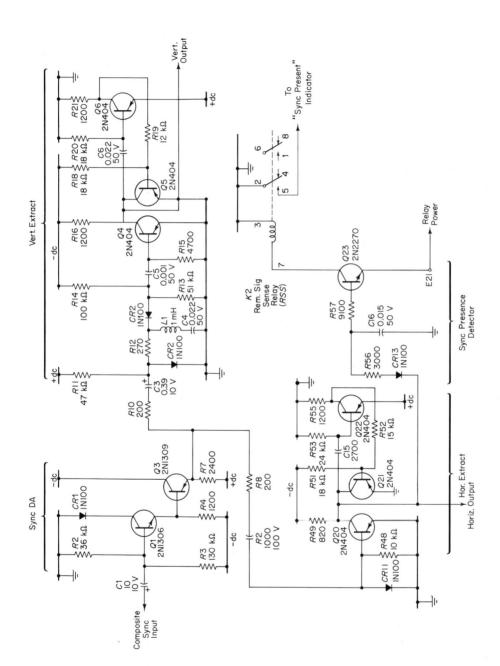

FIGURE 12-2. Novel sync separator drives sync lock circuits.

applied to $Q4$ that coincides with the leading edge of the second segment of vertical sync. This causes a positive-going pulse to be developed at the collector of $Q4$, which, in turn, triggers the monostable multivibrator consisting of $Q5$ and $Q6$. Thus, there is an output pulse occurring that coincides with the incoming vertical-sync interval, with the exception of a delay equal to the first vertical-sync pulse. Since this delay is constant and known, it is easy to compensate for it.

The horizontal sync, equalizing and serration information is coupled to the base of $Q20$ through $R8$ and $C2$. In order to eliminate twice-horizontal rate information occurring during equalizing pulse and vertical-sync time, the horizontal-sync extractor utilizes a multivibrator consisting of $Q21$ and $Q22$ to produce positive pulses that are three-quarters of a horizontal line in width, beginning with sync leading edges. The width of these pulses, being greater than a half-line, prevents the generation of pulses from those equalizing and vertical sync pulses that are not in step with the horizontal-sync pulses during the particular field being sampled. The output is derived from the collector of $Q20$, and an additional stage is incorporated in this particular circuit to trigger a relay which will light a lamp to indicate the presence of an incoming sync signal.

Vertical signal phase detection is accomplished by the circuitry shown in Figure 12-3. The phase detector is a sampling system in which a trapezoidal waveform (developed from the locally generated vertical information) is sampled by a pulse that was derived from the remotely generated vertical information. The diode bridge $Z1$ acts as a sampling switch. The trapezoidal waveform to be sampled is fed to terminals 3 and 8 of the bridge. A sampling pulse (obtained from the remote sync waveform) forces a pulse of current through the bridge from the 5 and 7 terminals to the 2 and 4 terminals. This biases all of the diodes in the bridge into the *on* condition, and effectively connects the 3 and 8 terminals to the output point at terminals 1 and 6. The instantaneous voltage appearing at the 3 and 8 terminals is thus caused to appear at the output. This voltage sample charges the storage capacitor, $C10$, which holds the charge between sampling pulses.

The voltage thus developed can be directed to a voltage-controlled oscillator in the sync generator to phase-lock the system so that the vertical output of the system will be in coincidence with that of the incoming sync waveform. If any drift is evidenced between the two, an error signal will be developed, changing the frequency of the oscillator until coincidence is once again established.

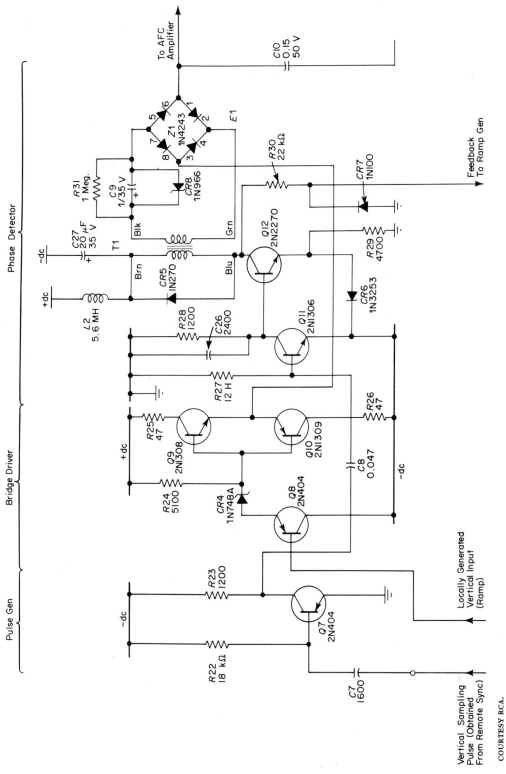

FIGURE 12-3. Phase detector compares remote vertical input with locally generated vertical signal and develops a dc difference signal.

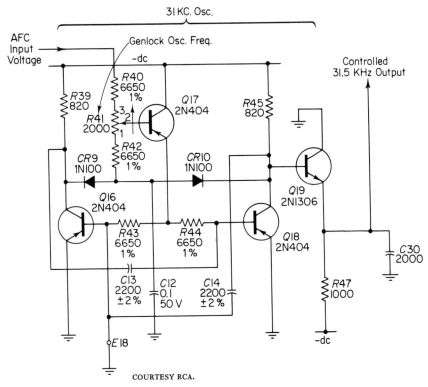

COURTESY RCA.

FIGURE 12-4. A voltage-controlled oscillator.

The dc voltage that was developed by the circuitry in Figure 12-3 is amplified and processed by additional circuits and applied to the voltage-controlled oscillator shown in Figure 12-4. This oscillator, a multivibrator consisting of $Q16$ and $Q18$, has a normal free-running frequency which is controlled by the values of the cross-coupling capacitors $C14$ and $C13$, and the developing resistors $R43$ and $R44$. However, transistor $Q17$ can be thought of as a variable resistance in series with the developing resistors $R43$ and $R44$. Causing the conduction characteristics of $Q17$ to vary will allow control of the frequency of oscillation, and it is to the base of $Q17$ that the AFC voltage developed in the previous circuit is applied. The 31.5-kilohertz output that results is then divided down to the vertical rate to provide the locally generated vertical information. Because this vertical information is that which is sampled by the previous phase detector, a servo loop has been established, and the vertical output will be coincident with the incoming vertical interval.

Horizontal sync may be sampled in much the same manner, and the output of its voltage-controlled oscillator may be divided by a factor of two to provide the horizontal-output information. If this is then compared to the incoming horizontal-sync waveform, a constant phase relationship may be established between the two.

## SUBCARRIER GENERATORS

The primary prerequisite of any subcarrier generator must be an extremely accurate and stable 3.579545-megahertz oscillator. Such oscillators are almost always crystal controlled, often contained in ovens to establish stable operation over wide temperature extremes, and they may also have an input terminal to provide for an AFC input voltage that will allow them to be locked to an external signal.

Figure 12-5 is a block diagram of an oscillator system which will operate in a crystal-locked mode, or use the burst signal from an external composite color signal to generate an AFC-voltage that will allow the oscillator to phase-lock to the burst sample.

In the diagram, the crystal oscillator drives a buffer amplifier which, in turn, provides input to a phase detector. The other input to the phase detector is a burst signal that has been removed from a composite color

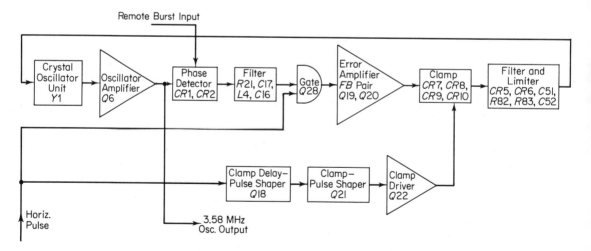

FIGURE 12-5. Subcarrier oscillator can phase-lock to incoming burst signal.

signal for comparison with the local oscillator frequency and phase. The detector establishes a dc-output potential that is a product of the phase difference between the two input signals and this filtered voltage is applied to a gate whose other input is a pulse which is coincident with the burst signal. The gate is therefore energized only during the interval when comparison is taking place and effectively excludes any other extraneous voltages. The error amplifier drives a clamp stage in which the error pulse output is clamped to ground. The resulting dc signal is filtered and applied to the oscillator as correction voltage. The oscillator changes phase according to the amplitude and polarity of the input, and adjusts itself until its output is in coincidence, both in frequency and phase, with the incoming burst waveform.

Figure 12-6 illustrates a color phase comparator. The two inputs, locally generated subcarrier and burst from the incoming signal, are compared and a dc voltage is generated that is a direct function of the phase difference between the two. The detector diodes, *CR*1 and *CR*2 are keyed into conduction by the amplified subcarrier oscillator signal, and after the first few cycles establish a charge on capacitor *C*13, the diodes conduct only during the signal peaks. Thus, each time the diodes conduct, the input line (containing the burst) is clamped to the charge established on the capacitor.

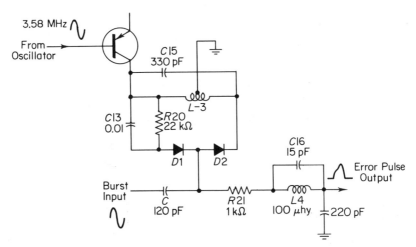

COURTESY COHU ELECTRONICS, INC.

FIGURE 12-6. Error pulse results from comparing burst input with locally generated color subcarrier.

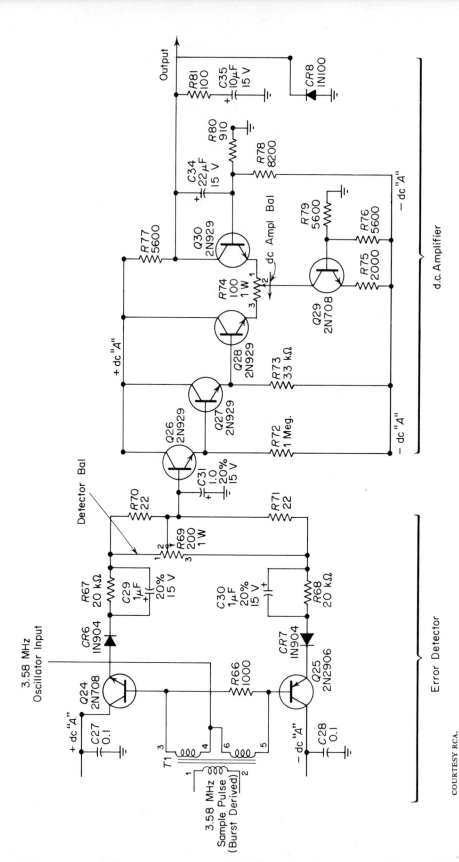

FIGURE 12-7. Another circuit yields dc output that represents phase difference between burst and local sub-carrier.

The burst is applied to the junction of the diodes and, when the diodes conduct for the short interval during the peak of the oscillator waveform, the incoming burst signal is clamped to the $C13$ potential. If the phase of the oscillator varies, the point on the burst waveform at which the circuit clamps will cause the burst to seek a different average dc level. Thus, if the phase relationship of the burst signal and the oscillator signal tends to shift, an error signal appears at the detector output in the form of an error pulse. The pulse amplitude will be a function of the phase difference. The error signal is filtered by $R21$, $C17$, $L4$, and $C16$, removing the 3.58-megahertz component. The error pulse that remains is amplified and clamped to ground, with the resulting dc voltage being filtered and applied to the oscillator as AFC control voltage, keeping the oscillator and the incoming burst in phase with each other.

Figure 12-7 is another example of a color phase comparator. Here the amplified and processed burst, which has been clipped to form an effective squarewave, is applied to the base of $Q23$, which acts as the sample pulse generator and drives the primary of $T1$. The locally generated subcarrier is applied to the center tap of $T1$ secondary. It should be noted that the two transistors $Q24$ and $Q25$ have their collectors returned to sources of different dc polarity. Thus, the conduction characteristics of these two transistors, when compared at the junction of $R70$ and $R71$, will determine the resulting dc voltage that is impressed upon the filter capacitor $C31$. The comparison of the burst and local oscillator frequency that takes place at $T1$ determines the conduction characteristics of $Q24$ and $Q25$, and the dc error voltage that is felt at the base of $Q26$ may then be amplified and used as AFC voltage for the local 3.58-megahertz oscillator.

## BURST-FLAG GENERATION

The burst of 8 or 9 cycles of subcarrier frequency that is contained on a composite color signal is placed there by a gating circuit that allows the subcarrier to pass only when a *burst-flag* pulse is present. This pulse is developed by pulse delay and shaping circuits that use the horizontal sync or blanking as drive. Figure 12-8 illustrates such a circuit.

Horizontal and vertical drive signals are applied at the input to gate $U12a$, which operates as a NOR gate. This NOR-gate output is a series of positive horizontal drive pulses which do not appear during the vertical serration or equalizing time of the composite sync waveform. These pulses

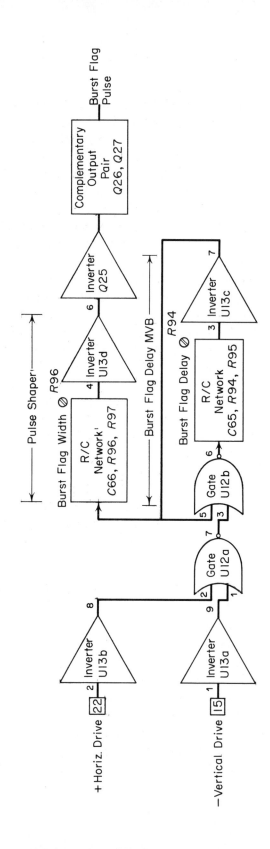

298

COURTESY COHU ELECTRONICS, INC.

FIGURE 12-8. Block diagram of burst-flag generator.

appear at the trigger input of a single-shot multivibrator comprised of *U*12*b,* *U*13*c,* and the *RC* network. *R*94 is adjusted to set multivibrator pulse width, which determines the "breezeway" (delay of burst with respect to the trailing edge of horizontal sync). The trailing edge of this multivibrator-output pulse initiates the tsart of the burst-flag pulse.

The multivibrator output is fed to a pulse-shaping circuit (*U*13*d,* *C*66, *R*96, and *R*97). *R*96 is adjusted to set the width of the burst-flag pulse. The width is adjusted so that it will be of proper duration to gate the desired number of cycles of subcarrier frequency onto the composite color waveform. The pulse-shaper output is inverted by *Q*25 and fed to complementary-symmetry output pair *Q*26 and *Q*27. The output here is 4 volts p-p negative into a 75-ohm load.

## ASSOCIATED EQUIPMENT— DOT-BAR GENERATOR

A dot-bar generator is often associated with sync generator equipment. It provides output waveforms which, when displayed on a monitor, generate either horizontal or vertical lines of limited width, a crosshatch or grating pattern consisting of both horizontal and vertical bars, or a series of dots which are made to occur (through gating procedures) at the point where the horizontal and vertical bars previously intersected.

A block diagram of a dot-bar generator is shown in Figure 12-9. Because all of the circuits are multivibrator, buffer, and boxcar circuits, with typical examples explained elsewhere in this book, no circuit configurations are shown.

The horizontal and vertical bar information is generated by sampling the input composite blanking information, separating it into its horizontal and vertical components and using it to establish start and stop times for the oscillators. One oscillator operates with a free-running period approximately twenty times that of the horizontal, or line, frequency. This circuit generates the vertical bars. The other oscillator generates the horizontal bars by operating at a frequency that is nominally about fifteen times the vertical, or field, frequency. Ahead of these oscillators are circuits which control the periodic starting and stopping of each oscillator. The vertical bar oscillator is started anew near the beginning of a horizontal line and stopped at the end of the line, and the horizontal bar oscillator is started near the beginning of each field and stopped at the end of each field. By

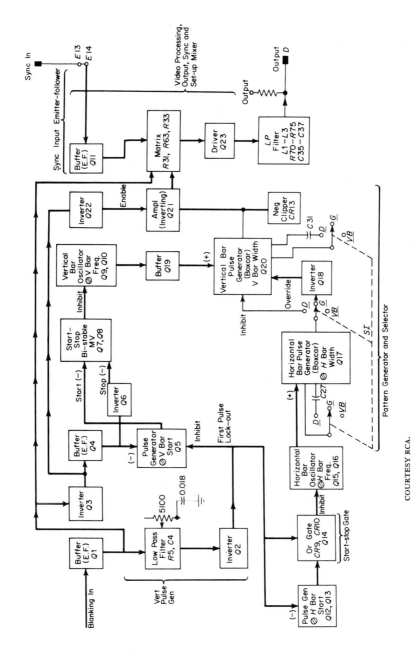

COURTESY RCA.

FIGURE 12-9. Block diagram of a dot-bar generator.

causing the oscillators to start and stop in this manner, the bars will be kept in phase with the scanning frequencies, and they will appear stable as viewed on the monitor presentation.

In operation, the blanking input is buffered by an emitter-follower amplifier $Q1$ and fed through a low-pass filter consisting of $R5$ and $C4$ which eliminates the horizontal pulses but allows the vertical blanking pulse to pass to the inverter block, $Q2$. The output of the inverter is an inverted vertical blanking pulse with slightly delayed edges. The output of the inverter is fed to the *horizontal bar start* block, a monostable multi-vibrator that generates a pulse starting with the trailing edge of the input vertical pulse. The pulse output from this stage is variable in width and it, coupled with the input vertical pulse, is used to control the starting time of the horizontal bar oscillator.

The vertical bar oscillator operates in a somewhat similar manner. The output from the buffer, $Q1$, is applied to inverter $Q3$ and is then routed to buffer $Q4$. The output from buffer $Q4$ feeds both the inverter $Q6$ and the pulse generator *vertical bar start* block, $Q5$. The combined action of these two stages develops triggering that causes rather wide pulses, occurring at the horizontal rate, at the output of the *start-stop* bistable multivibrator. This output pulse causes the vertical bar oscillator to start and stop at the proper intervals to produce stable vertical bars, and blank the output during retrace.

Proper matrixing of the two oscillator signals results in the ability to provide (in this case) vertical bars, a crosshatch or grating pattern, and a pattern of dots. The bars are useful for determining linearity characteristics and making quadrature adjustments associated with the headwheel of certain television tape machines. The crosshatch is used for linearity adjustments in camera and monitor deflection systems and for making convergence adjustments in color cameras. The dots are also useful in linearity and convergence procedures.

# 13

## COLOR ENCODERS AND
## ASSOCIATED EQUIPMENT

In the NTSC color system, color encoders have the task of converting the amplitude-modulated signals of the color camera into the $Y$, $I$, and $Q$ components of the composite color signal. The basic concepts and principles of this process have already been discussed in Chapter Ten. It remains, then, to examine the circuitry and equipment necessary to accomplish proper modulation and processing of the color signal.

Color encoders (also called *color modulators* or *Colorplexers*) may be an integral part of a color camera itself, or they may be a separate piece of equipment designed to work with many different types of cameras. Figure 13-1 shows a color encoder of the latter type.

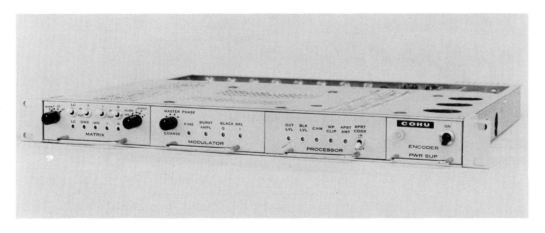

COURTESY COHU ELECTRONICS, INC.

FIGURE 13-1. A color video encoder.

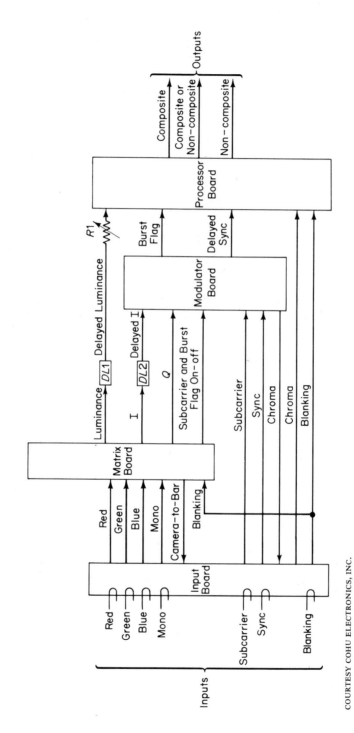

FIGURE 13-2. Color encoder block diagram.

## THE COLOR ENCODER

Figure 13-2 is a block diagram of a typical color encoder. It requires inputs of the red, green, and blue video voltages (and monochrome, in those cases where the camera is a four-tube device), subcarrier drive, and sync and blanking pulses. All of these are processed to produce the NTSC composite color signal.

All of the inputs shown in the example pass through an input assembly which provides isolation circuitry to reduce loading effects on the input and assure that the output signals are of the proper amplitude for use in the matrix, modulator, and processor sections of the subject system.

At the matrix section the red, green, and blue video from the input are matrixed and amplified to provide the *Y, I,* and *Q* video signals. The *I* and *Q* signals are amplitude-modulated waveforms at this point and are filtered and fed to the modulator. The filtering imparts the proper bandwidth characteristics to the *I* and *Q* signals, but it also causes the signals to experience some amount of delay. Therefore, before reaching the modulator board, the *I* signal is delayed by *DL2* to bring it into phase coincidence with the *Q* signal, which has experienced the greater delay due to its narrower bandwidth. The luminance signal that has been developed from the *R-G-B* inputs (or acquired as a separate input) must also be delayed, by *DL1*, so that it, too, maintains a proper phase relationship with the other components of the signal.

At the modulator board, the subcarrier signal is amplified and limited to obtain a constant amplitude. This signal is applied directly to the *I* modulator and is also delayed by 90 degrees and applied to the *Q* (quadrature) modulator. The *I* and *Q* signals are also routed to their respective modulators and the sideband-modulated *I* and *Q* signals are realized. The composite sync signal is delayed on the modulator assembly to keep it coincident with the luminance and chrominance information.

The processor block is, as its name indicates, that portion of the circuitry where the composite signal is amplified, peaked, and processed. Blanking and burst are added and the chrominance information is summed with the luminance signal to form the NTSC composite color signal required for broadcasting applications.

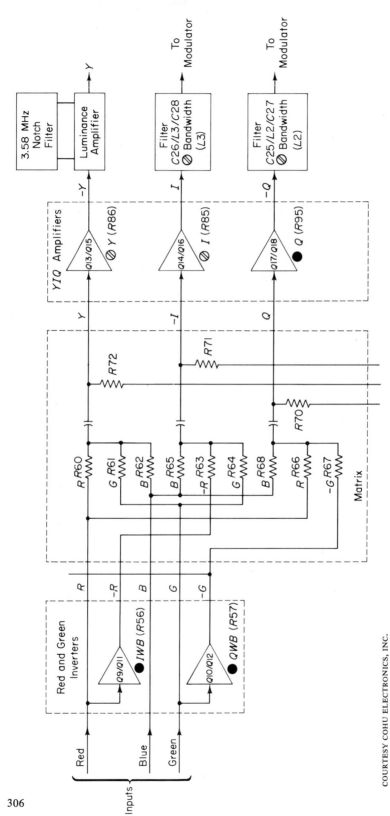

COURTESY COHU ELECTRONICS, INC.

FIGURE 13-3. $Y$, $I$, and $Q$ matrix.

## The Matrix Circuits

Figure 13-3 illustrates a block diagram of the matrix network needed to convert the *R, G,* and *B* input-voltage waveforms to the *I, Q,* and *Y* signals needed for proper signal content. The red and green signals are inverted before being applied to certain sections of the matrix network. Thus, the matrix-input signals consist of *R, −R, B, G,* and *−G.* Each resistor receives only one input, and the output of each of the three resistor networks is capacitively coupled to the *Y, I,* and *Q* amplifiers. Each set of resistors is ratioed so that the currents in the resistors are summed to obtain the following standard signal proportions:

$$Y = 30\%\,R + 59\%\,G + 11\%\,B$$
$$-I = -60\%\,R + 28\%\,G + 32\%\,B$$
$$Q = 21\%\,R - 52\%\,G + 31\%\,B$$

The *Y* equation is, of course, derived from the proportionate values of red, blue, and green needed to create a 100 per cent brightness characteristic of white. The *I* and *Q* equations are such that, when added together, their vector sum will correctly represent the color content of the signal. It may be remembered that the resultant phase carries hue information and the amplitude carries the saturation information.

In the event that a monochrome, or luminance signal, is available from the camera, the *Y* output may be switched from the matrix result to the camera *Y* signal output.

## The Modulator Circuits

The modulator circuits must modulate the 3.58-megahertz subcarrier with the *I* and *Q* signals. Figure 13-4 illustrates a block diagram of the *I* (in-phase) and *Q* (quadrature) modulators. Both are identical in circuitry and operation except for the 90-degree phase shifter that is incorporated at the input of the *Q* modulator to place the *Q* reference vector at right angles to the *I* reference. The components shown in the phase-shifter network will allow a precise 90-degree phase shift at the 3.58-megahertz subcarrier frequency. A trimming adjustment is provided by *L*4 to facilitate critical alignment.

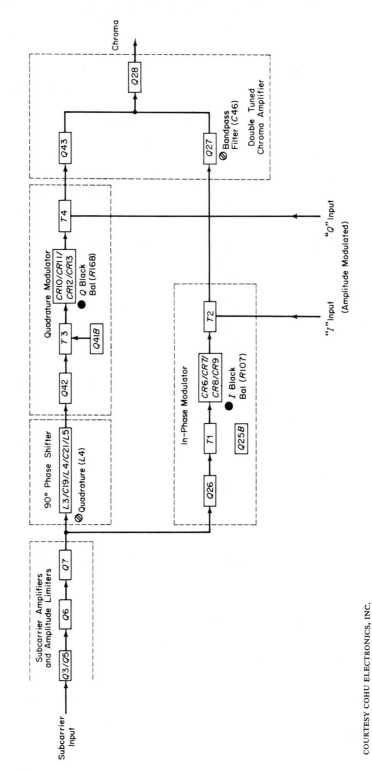

COURTESY COHU ELECTRONICS, INC.

FIGURE 13-4. Q modulator employs 90-degree phase shift at subcarrier input.

Figure 13-5 schematically illustrates the configuration of the double-balanced modulators indicated in the block diagram. $Q26$ acts as an emitter follower to drive the primary winding of $T1$ with the 3.58-megahertz subcarrier. The other input, the amplitude-modulated signal from the $I$ amplifier, is applied to a center tap on the primary of $T2$, with a return path via a center tap on the secondary winding of $T1$. The 3.58 megahertz is felt at the secondary of $T1$, with each end of the transformer producing a subcarrier signal that is 180 degrees out-of-phase with that of the signal at the other end. These subcarrier signals are applied to the diode bridge consisting of $CR6$, $CR7$, $CR8$, and $CR9$. However, the video applied at

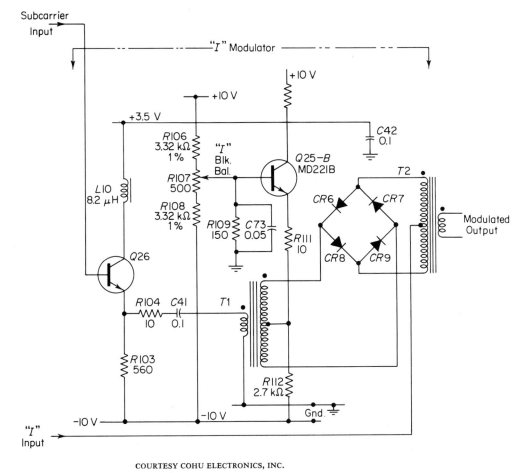

FIGURE 13-5. A double-balanced modulator encodes the $I$ signal.

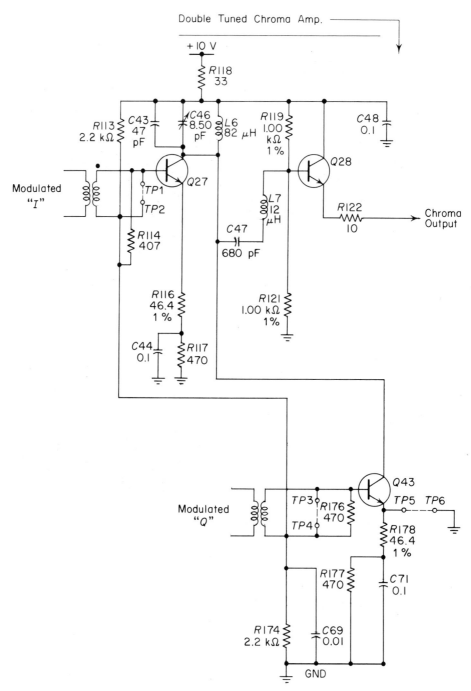

FIGURE 13-6. Double-tuned amplifier combines modulated $I$ and $Q$ signals.

the center tap of the $T2$ transformer is felt as the same polarity at both ends of that winding, and it is used to bias the diodes, thus affecting the configuration of the 3.58-megahertz output.

If the video ($I$ or $Q$) is positive, it will be felt as such at the diode bridge, and diodes $CR6$ and $CR9$ will be biased *on,* routing the 3.58 megahertz to the primary of $T2$ in such a way that the top end of $T1$ is connected to the top end of $T2$. If the video is negative, it will cause diodes $CR7$ and $CR8$ to conduct, and the 3.58 megahertz will be fed to the primary of $T2$ through their forward-biased junctions, this time with the bottom end of $T1$ connected to the top of $T2$ through $CR7$ and the top of $T1$ connected to the bottom of $T2$ through $CR8$. Therefore, these two states of conduction, resulting from a positive or negative input signal ($I$ or $Q$) at the center tap of $T2$, will result in two output signals that are 180 degrees out-of-phase with each other. A positive $Q$ signal input, for instance, will give a 3.58-megahertz subcarrier output that is 180 degrees out-of-phase with that which would result if a negative $Q$ signal were applied. If no input is applied, the diode bridge will not be biased into an *on* condition, and there will be no subcarrier output from the modulator. Thus, when no color information is present, there will be no color sideband output.

The chrominance information obtained from the illustrated $I$ and $Q$ modulators is summed in a double-tuned chroma amplifier, as shown in Figure 13-6. The circuit consists of tuned amplifiers $Q27$ and $Q43$ and emitter follower $Q28$. The $Q27$ and $Q43$ circuits amplify the outputs of the $I$ and $Q$ modulators, respectively. The amplifiers have a common load which is tuned by variable capacitor, $C46$. The combined outputs are thus summed to form the resultant color vector described in Chapter Ten.

When several color cameras are in use, it is very important that all of the color signals bear the same phase relationship with each other. For this reason, a single subcarrier oscillator source is generally used to drive all cameras. However, because the cable lengths to the different cameras may not be the same, it is necessary to incorporate a means whereby the phase of the subcarrier input to each encoder system can be adjusted.

Figure 13-7 illustrates a phase shift network that can provide a 360-degree phase shift at the 3.58-megahertz subcarrier frequency before it is applied to the modulators. The subcarrier input associated with each individual camera system can then be adjusted so that all modulators will be subcarrier at the same phase. (The $Q$ modulator will, of course, receive

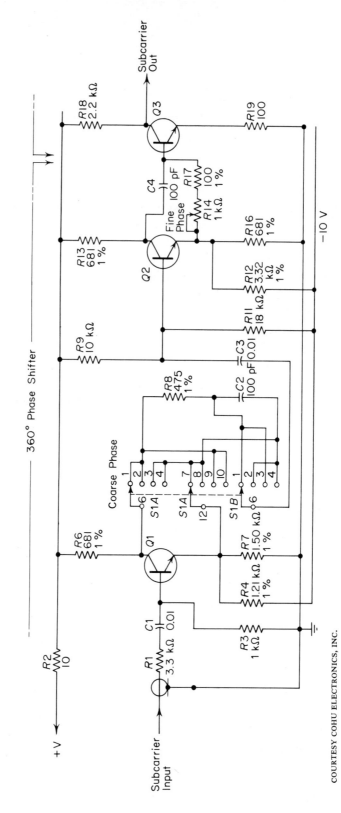

COURTESY COHU ELECTRONICS, INC.

FIGURE 13-7. Circuit provides 360-degree phase shift capability at subcarrier frequency.

subcarrier that has been shifted an additional 90 degrees, as previously shown).

In the diagram, the output of $Q1$, a phase splitter, is routed through a phasing and switching network to a second-phase splitter, $Q2$. $Q1$ allows selection of an in-phase subcarrier at its emitter, or a 180-degree out-of-phase signal at its collector. An additional phase shift capability is provided by switching capacitor $C2$ in or out of the circuit. These course adjustments are complemented by the fine phase adjustment provided by potentiometer $R14$.

### Color Video Processing

Processing of the monochrome portion of the color video signal is quite similar to that shown in Chapter Six. There must be consideration given to clamping circuits, aperture correction, blanking insertion, and the like. One noteworthy difference in the video processing circuitry is the summing network necessary to combine all of the various signals into one composite color signal output. This circuit is schematically illustrated in Figure 13-8.

In the diagram, the chroma information is applied to the summing transistor $Q14$ through adjustable-frequency harmonic filters $L3/C34$, $L4/C38$, and $L5/C40$. Luminance signal is applied at the emitter of $Q14$, and the clamping action necessary to provide a stable reference level is achieved by the driven diode bridge, $CR4$, $CR5$, $CR6$, and $CR7$. The output transistors, $Q17$, $Q18$, and $Q19$ form a potentiometric-feedback amplifier. The output stage, $Q18/Q19$ is a unity-gain feedback amplifier which provides a very low output impedance. The output is distributed through three 75-ohm, source-terminating resistors. It may be seen that sync is added to the output at terminal 5, may be selected *on* or *off* at terminal 7, and is not present at terminal 9. Thus, of the three outputs, one is composite, one noncomposite, and the other is selectable.

## COLOR-BAR GENERATOR

The purpose of a color-bar generator is to provide accurate red-green-blue and $I$-$Y$-$Q$ color-bar signals to test, calibrate, and troubleshoot monitors, encoders, and other studio or closed circuit color equipment.

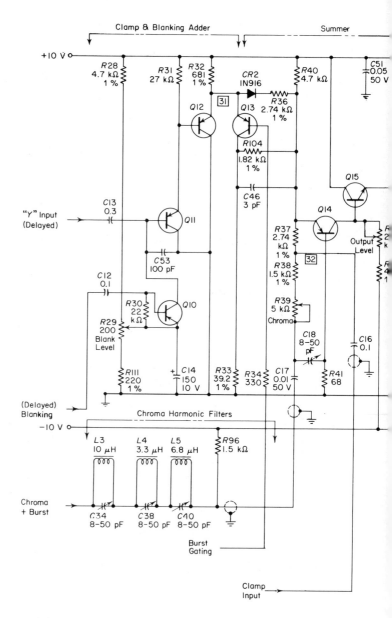

FIGURE 13-8. Composite signal generation.

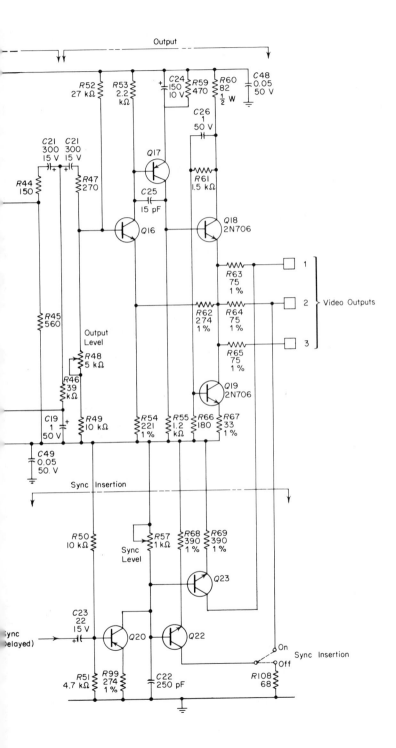

Two monitor presentations and their associated color video signals are shown in Figure 13-9. These demonstrate two standard output signals that may be obtained from a typical color-bar generator. If the amplitude-modulated signals that generate these colors are accurately controlled as to their amplitude, the result may be used to adjust the phasing of a color

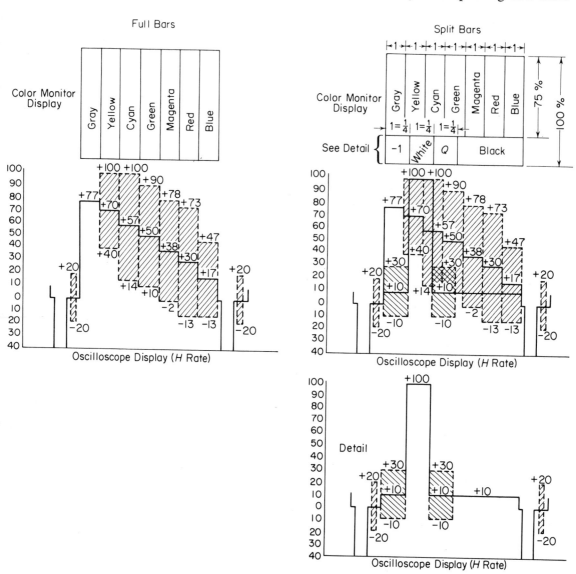

FIGURE 13-9. Color bars and their video signals.

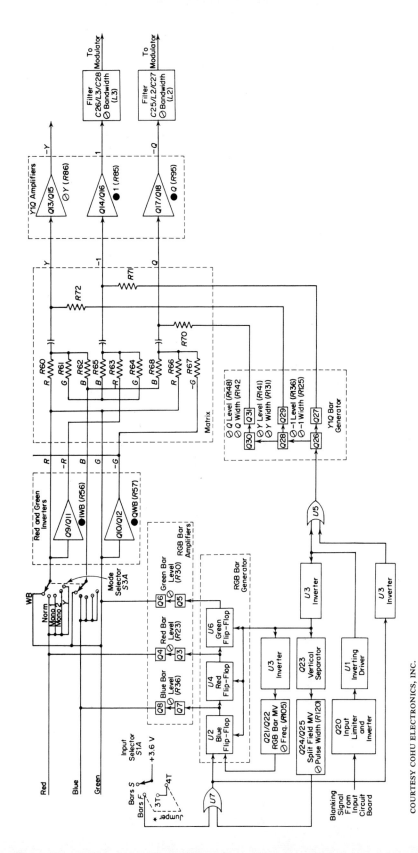

FIGURE 13-10. A color-bar generator provides test signals.

modulator. This may be done by use of a *vectorscope,* which displays phasing as a vector displacement on the cathode-ray tube. The *R-G-B* outputs of a color-bar generator, may then be used as input to a color encoder, instead of the camera signals, and very accurate adjustments to the encoder can be performed.

Figure 13-10 illustrates a color-bar generator connected to a matrix network. The bars are generated by multivibrators that are triggered by the horizontal blanking signal from the system sync generator. It can be seen that the *R-G-B* outputs are applied at the same inputs as the camera inputs. In the "full bars" configuration, each horizontal interval would contain the *R-G-B* video waveforms shown in Figure 13-11. During the first interval, the addition of the equal-amplitude *R, G,* and *B* results in a gray output. The next bar, yellow, is the product of a green and red output, with no output being realized from the blue channel. Cyan is the result of green and blue outputs and an absence of red signal. The other color bars are similarly composed.

When a split-screen effect is desired, the *I, Y,* and *Q* signals, followed by a black interval, are displayed on the lower half of the presentation.

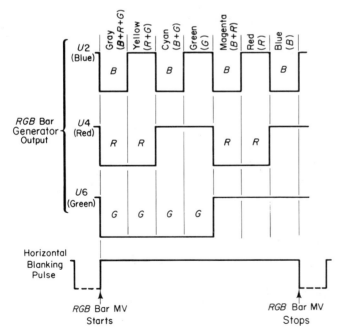

FIGURE 13-11. *RGB* outputs from a color-bar generator.

This is accomplished by routing equal-amplitude pulses, in a sequential manner, to the $I$, $Y$, and $Q$ outputs without first going through the matrix. Thus, there is developed a signal output first in the $I$ channel, with no luminance or $Q$ components present. This is followed by a 100-per cent white bar, with no $I$ or $Q$ components being developed and, finally, a $Q$ output is generated.

The ability to generate the above signals makes it much simpler to align and maintain equipment that generates and uses the NTSC composite color signal.

# COLOR RECEIVERS
# AND MONITORS

Color television monitors assume two general configurations—those which accept three different, unmodulated signals to represent the red, green, and blue color information, and those which use the modulated NTSC-type color video signal to achieve a color presentation. The first is commonly known as an *R-G-B* monitor, and the second as a *decoding* color monitor.

The color television receiver generally differs from a closed circuit color monitor only in that it incorporates RF and IF sections to enable it to receive broadcast transmissions. In addition, a sound section is included which is normally not a part of a closed circuit monitor. Aside from these features, additional circuits, such as the deflection systems, video processors and amplifiers, chroma demodulators, etc., provide essentially the same functions.

Although no attempt will be made to give extensive analysis of receiver circuits, a familiarity with the techniques involved in the reception and processing of RF signals is certainly desirable. For this reason, several examples of circuitry found in the RF tuner and IF sections of a modern solid-state receiver are also discussed.

## THE RF TUNER

It is the purpose of the RF tuner to receive and amplify the RF signals broadcast by television stations, and provide a means whereby individual stations can be selected for viewing according to their characteristic frequencies. Figure 14-1 shows a basic block diagram of the RF tuner. The RF-amplifier section increases the amplitude of the signal received by the antenna and generally provides selective filtering of unwanted frequencies by means of filter networks which are inserted as the channel selector is moved from one channel position to another. The output of the RF amplifier is fed into a mixer where it is combined with the signal being generated by the oscillator section. The oscillator frequency is controlled by the position of the channel selector and is chosen so that the mixer output, which is a product of the selected RF signal and the oscillator frequency, constitutes an intermediate frequency of approximately 45 megahertz.

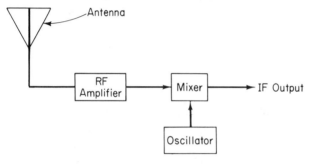

FIGURE 14-1. Block diagram of television tuner.

In the color receiver, the requirements of the RF tuner are more stringent than those used in monochrome receivers. Each channel position must present a flat response over the entire band so the chrominance information is not attenuated. In addition, stability of the oscillator is a very important factor in the reception of color signals.

For purposes of explanation, it might be well initially to discuss a tuner designed for use in a monochrome receiver. It is somewhat more simple in that it lacks the bandwidth demands of the color tuner. Figure 14-2 is a schematic illustration of such a tuner.

$Q1$ acts as the RF amplifier and receives its input from the antenna through the input capacitor, $C3$. Frequency rejection of unwanted signals is accomplished by the action of the tank circuit that is connected from the input side of $C3$ to ground. The tank circuit is formed of $C29$ and a value of inductance which is selected by positioning the channel selector.

The oscillator, $Q3$, runs at a frequency determined by the value of inductance which is switched into the circuit by the channel selector. $L29$ through $L40$ are wound for channels 2 through 13. Fine tuning is individ-

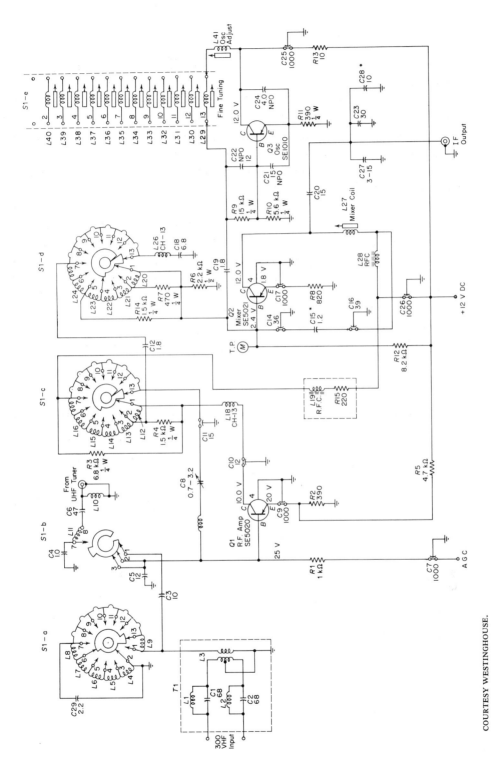

FIGURE 14-2. A VHF tuner.

ually adjustable for each channel by means of a variable slug in each of the coils.

The oscillator output is coupled by $C19$ to the base of mixer transistor $Q2$ where it is combined with the RF signal from $Q1$. The $Q1$ signal finds its way to $Q2$ through the various inductive and capacitive elements that are a part of the channel selector. The product of the two signals is an intermediate-frequency signal of approximately 45 megahertz that contains the video signal. This output is routed to the IF stages for further processing before the video detection takes place.

Another feature of the tuner is the AGC voltage that is applied to the base of the RF amplifier, $Q1$. When the video is detected in later stages of the receiver, a dc potential is developed that is proportional to the strength of the received signal. This voltage is used to bias $Q1$ so that its effective gain is a function of the AGC voltage. Thus, an automatic gain control is achieved which keeps the video signal relatively constant regardless of the signal strength of the incoming RF.

When it is desired to tune to a UHF station, the channel selector in

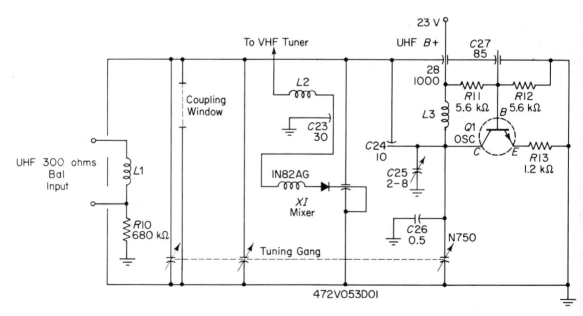

Note.
1. All Capacitance Values in pF.

COURTESY WESTINGHOUSE.

FIGURE 14-3. A UHF tuner.

most receivers is positioned between channels 13 and 2. In this position, the tuner in Figure 14-2 uses the S-1b section of the tuner switch to disconnect the VHF antenna and coupling elements and connect the output from the UHF tuner to the base of $Q1$. The VHF oscillator is also disabled because no coil is provided for its use with the tuner in the UHF position. The output from the UHF tuner is therefore amplified and routed through $Q2$ directly to the IF section.

Figure 14-3 is a schematic of a UHF tuner which will provide an output at the required 45-megahertz IF frequency. $Q1$ forms the oscillator section, and its frequency is continuously variable by use of one section of the variable capacitors that form the tuning control. Diode $X1$ acts as the mixer stage, and provides a resultant IF output of approximately 44 to 45 megahertz. No direct connections are made from the antenna input, $L1$, to the mixer stage because RF coupling is used.

Figure 14-4 illustrates the circuitry of a VHF tuner that was designed for use in a color television receiver. The configuration here is somewhat more complex than in the previous example because of the demands of the color signal; however, the operational theory is essentially the same. Those components marked "CH" are selectable, depending upon the position of the channel selector, and they fulfill the role of the wafer switches in the previous example.

## THE VIDEO IF SECTION

The intermediate-frequency amplifiers serve to amplify the output of the RF tuner and provide a bandpass that will attenuate those frequencies that lie outside the area of interest. These requirements apply equally for monochrome and color receivers although, again, the color receiver is somewhat more demanding because the IF section must maintain a flat-frequency response to the high-frequency end of the video for proper processing of the chrominance information.

In a color receiver provision must also be made for the sound subcarrier to be attenuated at some point before the video detector stage. The sound subcarrier must not be present in the detected video signal or a beat signal will be produced between the 4.5-megahertz sound subcarrier frequency and the color subcarrier of 3.58 megahertz. The beat signal of approximately 920 kilohertz that would be produced would result in objectionable interference patterns on the picture tube.

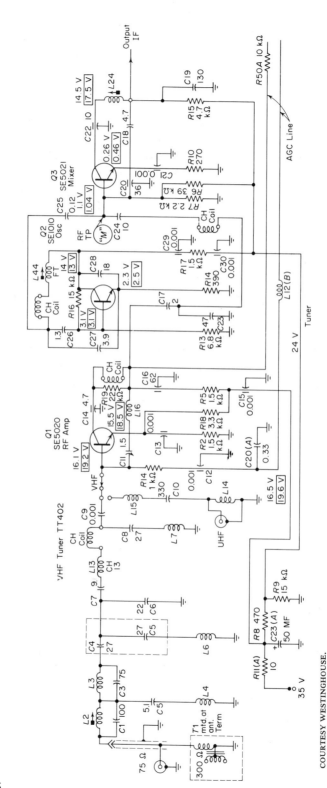

COURTESY WESTINGHOUSE.

FIGURE 14-4. A VHF tuner for color TV.

Figure 14-5 is a simplified block diagram of an IF section and the detectors which follow. It can be seen that an AGC voltage is applied to the IF circuitry, as well as the RF amplifier in the tuner. The AGC is applied to the base of the second video IF amplifier to control its relative gain. The dc voltage that results at the emitter of the second amplifier is routed to the base of the first IF amplifier as a gain control. Thus, both the first and second stages of IF amplification are gain controlled by the application of a variable dc potential to the base of the second amplifier.

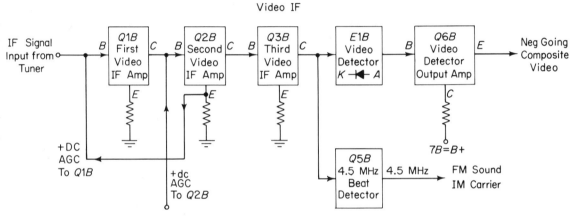

COURTESY WESTINGHOUSE.

FIGURE 14-5. Video IF employs AGC.

The output of the third IF amplifier is fed into the video detector and the 4.5-megahertz beat detector. However, filtering is performed in the video detector to eliminate any 4.5 megahertz that might be present.

Figure 14-6 shows the schematic of the IF section under discussion.

The input to the first amplifier, $Q1$, contains a 47.25-megahertz trap to eliminate interference from adjacent channel sound, and a 39.75-megahertz trap to reduce interference from adjacent channel video. Interstage coupling between the three amplifiers is accomplished by use of the familiar IF transformers.

Diode $E1$ acts as the video detector, and particular note should be taken of the 41.25-megahertz trap in this circuit. The trap eliminates the sound subcarrier and the video will therefore contain no 4.5 megahertz to beat with the 3.58-megahertz chrominance. The IF output to the 4.5-megahertz detector is taken directly from the collector of $Q3$ before any filtering of the 41.25 megahertz takes place.

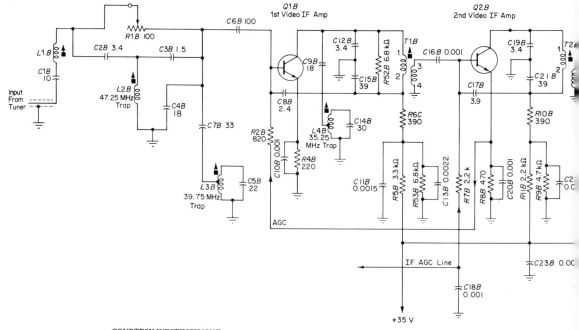

COURTESY WESTINGHOUSE.

FIGURE 14-6. Schematic of Figure 14-5.

The capacitive networks in the collector circuits of the first two IF stages, and the small capacitor that connects the bottom end of the transformer primary to the base of the driving transistor act to neutralize the stage. This effectively makes each transistor strictly a one-way device, preventing output signals from becoming a part of the input.

## KEYED AGC

The development of a voltage that will control the gain of the RF amplifier and the IF section in proportion to the amount of signal that is received is a function of the "keyed" AGC stage. Figure 14-7 illustrates a circuit that will develop such a voltage for automatic gain control.

Two inputs are required for proper operation of the AGC control. Because it is desirable to sample a portion of the input signal that is not subject to change with picture content, a sample of the signal is taken during the horizontal retrace period. Therefore, that portion of the signal that is sampled, namely the sync signal, will vary only as the signal strength

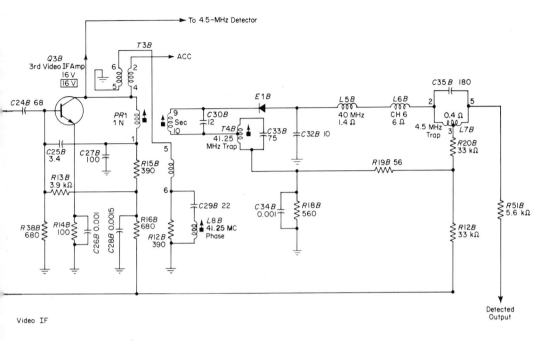

To 4.5-MHz Detector

ACC

*Video IF*

Detected Output

of the tuner input varies, and the gain will be adjusted to compensate for the variation. The gain of the receiver will thus be automatically adjusted when switching between channels that are associated with signals of widely varying strength.

In operation, transistor $Q300$ is normally cut off due to the bias applied to its base. Two input signals are applied to the base, a pulse at the horizontal rate and video from the video amplifier stages, both of which are negative going in polarity. As shown, diode $X301$ is back biased with the negative video information applied to its anode. The diode will therefore conduct only during the interval that the flyback pulse is present at the cathode of the diode. During this period, two events take place: the diode becomes forward biased and allows the video signal to pass, and the transistor becomes forward biased and conducts.

When the transistor conducts, a signal is developed at point $A$ that is a function of the current flow through the transistor. Since this is directly related to the video signal at the input when the transistor is keyed on, the signal amplitude that is developed is directly related to the video amplitude. Capacitor $C301$ filters the signal at $A$ into a dc voltage that is positive in potential. Because the voltage divider at the anode of diode $X300$ creates a positive voltage also, the diode is normally biased *on*. Thus, $R306$ is in parallel with the AGC *crossover* control and $R304$. On low signal levels, the IF AGC voltage will be approximately two volts. As the signal

329

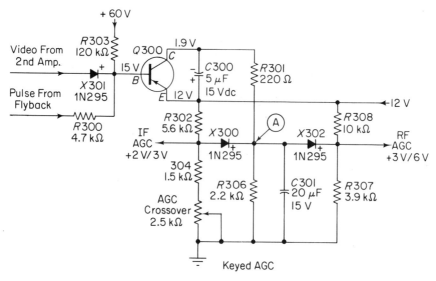

COURTESY WESTINGHOUSE.

FIGURE 14-7. Automatic gain control samples video.

strength increases, the voltage at point $A$ increases and the IF AGC voltage rises in proportion. This is true up to the point where diode $X300$ becomes reverse biased, and at this point the AGC voltage remains constant even if signal strength increases still further.

The AGC to the RF stage is developed similarly by diode $X302$. At low signal levels the diode is reverse biased. When increasing video levels increase the potential at $A$ to a point where $X302$ begins to conduct, the AGC voltage to the RF stage will begin to increase. This increase will reduce the gain of the RF amplifier, thus adjusting the signal level.

It could normally be supposed that such positive voltage developed by an AGC circuit would be fed to p-n-p transistors in the RF and IF stages to reduce gain as voltage level increases. However, this may not be always be the case, as evidenced by Figure 14-8A. The simplified schematic of two n-p-n IF stages would lead one to believe that a reduction in gain could be caused only by providing a more negative voltage at the base of $Q201$, reducing conduction. Nevertheless, attention to the graph in Figure 14-8B shows that the transistors used in this particular application have some very special characteristics. As bias is increased, collector current increases in the normal manner, but ac gain *decreases*. This is known as *forward AGC* and is accomplished only because of the internal composition of the specially designed transistors. The gain does rise normally to a certain point, but beyond this area an increase in collector current will reduce the ac gain. For comparison, the gain curve of a normal transistor is also shown.

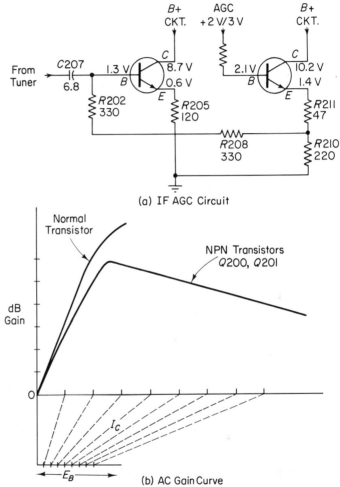

(a) IF AGC Circuit

(b) AC Gain Curve

FIGURE 14-8. IF AGC circuit employs special transistors.

## THE LUMINANCE SIGNAL

It has previously been demonstrated that the luminance signal in color television is identical to the video signal commonly associated with monochrome television, and it is processed in a similar manner. Figure 14-9 shows a simplified flow path for the luminance signal, together with the

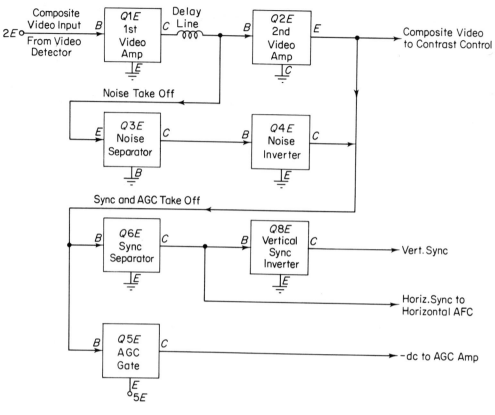

COURTESY MOTOROLA.

FIGURE 14-9. Luminance signal is delayed to compensate for color signal delay.

associated noise circuits and sync and AGC blocks. The delay line is included to compensate for the delay that the chrominance signal experiences in the bandwidth-limited chroma circuits.

The circuitry comprising the two video amplifiers and delay line of the diagram is shown schematically in Figure 14-10. Transistor $Q1$ acts as a standard common-emitter amplifier, providing drive through the delay line to $Q2$, an emitter follower whose output contains a variable-amplitude takeoff potentiometer which comprises the contrast control. The video from this point is routed to the three final drive amplifiers that provide drive to the individual cathodes of the tricolor picture tube. The noise takeoff shown in both the block diagram and schematic routes a video sample to the noise separator where the noise is removed from the video and inverted in the noise inverter. The inverted noise is then added to the emitter of $Q2$ via the takeoff line connected to that point, where it tends to cancel the original noise. The signals used to drive the sync separator and AGC circuits are also taken from the emitter of $Q2$.

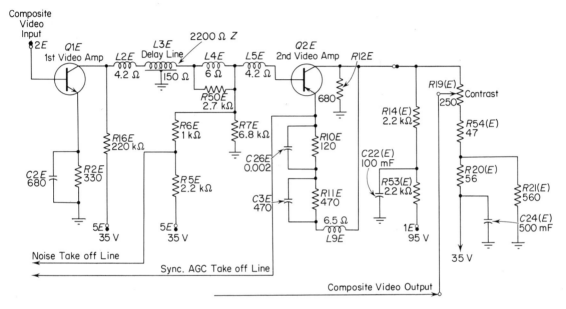

COURTESY MOTOROLA.

FIGURE 14-10. Circuitry of Figure 14-9.

## NOISE CANCELLATION

A somewhat different circuit arrangement that demonstrates the principles of noise cancellation appears in Figure 14-11A. Here the first video-amplifier stage is shunted by the noise cancel circuit and the effect is to remove and invert only the noise pulses of extreme amplitude. These pulses are then coupled back to the video at the emitter of the amplifier where they mix with and cancel the original noise pulses. Figure 14-11B illustrates the actual noise cancelling circuit whereby the transistor is connected as a simple inverting amplifier that is normally cut off due to the reverse bias between the base and emitter. The voltage at the base also keeps diode $X304$ reverse biased. Therefore, the circuit is effectively inoperative until a negative noise-pulse of sufficient amplitude occurs which will over-come the bias on the diode and trigger the transistor into the *on* state. The transistor remains on only for the duration of the noise pulse, and a signal appears at the collector which is coincident with, but opposite in polarity to, the original noise. This signal is then added to the original video where it tends to cancel the noise spikes.

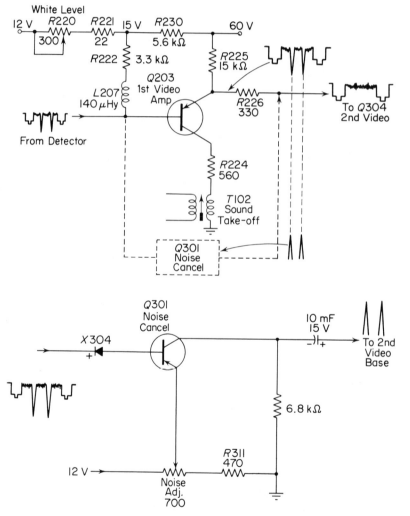

FIGURE 14-11. Noise cancelling circuit employs inverter stage.

The *noise adjustment* control is normally adjusted to a point just prior to where the stage will trigger on the negative-going sync pulses. Thus, any noise pulse which exceeds the amplitude of the sync pulse will be eliminated by cancellation. This not only reduces the obvious visual disturbance that would be caused, but also eliminates a pulse that could cause disruption in the sync separator circuits.

## CHROMA DEMODULATORS

There are many types of demodulator circuits which can be used to remove the color information from the composite color signal. Because of the nature of the color signal, all such circuits must be sensitive to a change in the phase of the input signal, giving an output signal that varies in amplitude according to the amount of phase shift. Any such phase shift is detected by comparing the chrominance information with a locally generated 3.58-megahertz signal that is phase-locked to the burst signal on the composite video waveform. If the reproduced colors are to be an accurate representation of the color camera output signal, it is imperative that the local 3.58 megahertz maintain extremely close phase tolerances with the burst, since this is the reference upon which all of the color information is founded.

It may be remembered that the $I$ and $Q$ signals that were generated in the camera system were made possible by using two 3.58-megahertz signals that were 90 degrees out-of-phase with each other. Combining these two signals in varying amounts of amplitude achieved the resultant signal that is referred to as the chrominance portion of the composite color video waveform. Demodulation of this resultant signal is achieved by again using two signals that are 90 degrees out-of-phase with each other.

Figure 14-12 illustrates a simple block diagram of a chroma demodulator. In this case the chrominance information is fed into two demodulator stages, one of which represents the $R-Y$ output and the other the $B-Y$ output. The $B-Y$ input is delayed by a period of time corresponding to 90 degrees of the chrominance frequency.

The locally generated 3.58 megahertz is also applied to the two stages. The output of the demodulators is a product of the phase difference between the two inputs. The output is devoid of the 3.58-megahertz subcarrier frequency, because it is effectively filtered from the resultant video. From this point on the color information is an amplitude-modulated signal.

Two such amplitude-modulated signals are obtained from the demodulators. These, the *R-Y* and *B-Y,* are matrixed together to obtain the third necessary signal, the *G-Y*. Once these three components of the color signal are reproduced, it is simply a matter of adding the luminance, or *Y,* signal to each to obtain the red, blue, and green signals that were originally generated by the three respective camera tubes in the color camera.

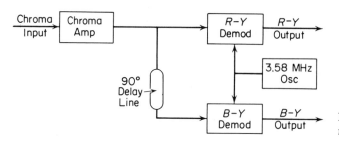

FIGURE 14-12. Chroma to *B-Y* demodulator is delayed by 90 degrees.

It should be pointed out here that most modern receivers and monitors are designed to recover the *R-Y* and *B-Y* signals instead of the *I* and *Q* signals that are transmitted. This is quite easily accomplished because the only difference between the *I* and *Q* and the *R-Y* and *B-Y* signals is a phase difference of approximately 33 degrees on both axes. When demodulated, the difference is reflected in the following equations:

$$I = 0.74(R\text{-}Y) + 0.41(B\text{-}Y)$$
$$Q = 0.48(R\text{-}Y) + 0.41(B\text{-}Y)$$

The demodulated color signal is referenced to an axis that is determined by the phase of the locally generated subcarrier. Any shift or change in the subcarrier will change the resultant color of the detected signal. Thus, if a control is provided that allows the local subcarrier to be shifted in phase with respect to the burst signal to which it is referenced, this will provide a means whereby the axis may be shifted.

Once the subcarrier phase has been established, the demodulator uses it as a standard for comparison with the chrominance signal.

Figure 14-13 illustrates a simple type of demodulator that uses a multi-element vacuum tube as the comparator. Although we are concerned primarily with solid-state devices, analysis of this circuit will lend understanding of the basic principles involved with phase demodulation.

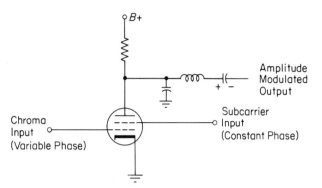

COURTESY WESTINGHOUSE.

FIGURE 14-13. Demodulator output amplitude is dependent upon input phase.

As previously stated, the purpose of the demodulator is to convert the phase and amplitude differences of the input signal into a variable amplitude signal which can be used to drive the color picture tube. In this example, only one channel, either the *R-Y* or *B-Y,* is illustrated.

Figure 14-14A illustrates the average output level with only the reference subcarrier applied. This *G*3 signal results in an average plate voltage which is shown to the right of the diagram. The 3.58-megahertz component is filtered out by the action of *L*504 and *C*518. In operation, the tube conducts during the positive half-cycle of the input signal and is cut off during the negative half-cycle. When an input signal that is in phase with the subcarrier is applied to *G*1, the collector current increases during the positive half-cycles, and the average plate voltage decreases (Figure 14-14B).

When the input signal is 180 degrees out-of-phase with the signal at *G*3, in Figure 14-14C, the amount of collector current decreases because of the opposing drive voltages and the average plate voltage increases accordingly.

In Figure 14-14D, it may be noted that a signal at *G*1 that is 90 degrees out-of-phase with that at *G*3 will result in an output which is identical with that which occurs when no input is applied at *G*1.

The above example illustrates a relatively simple means of converting a phase-modulated signal into amplitude excursions that are suitable as CRT drive signals when properly amplified.

Figure 14-15 shows a demodulator scheme that relies on diodes to demodulate the chrominance signal. Its principle of operation is somewhat different from that shown in Figure 14-14, but the end result is the same.

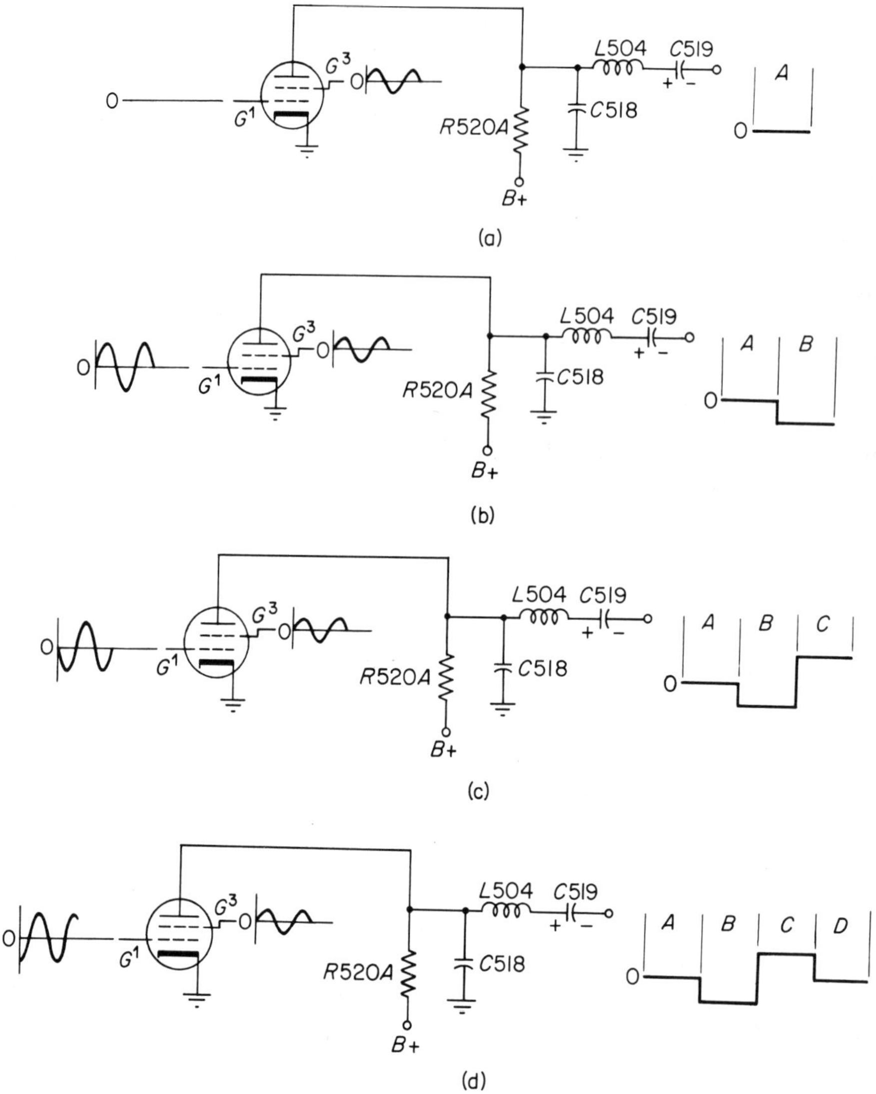

COURTESY WESTINGHOUSE.

FIGURE 14-14. Output amplitude varies with input phase shift.

The diagram is drawn in two different configurations. Figure 14-15A is the conventional schematic and 14-15B is an equivalent drawing that lends itself to easier understanding.

With no signal applied to the input, the two diodes will act in a manner somewhat similar to a simple power supply, that is, they will charge

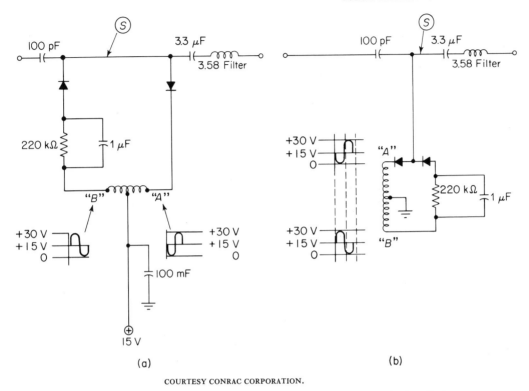

FIGURE 14-15. A chroma demodulator employing diodes.

capacitor $C1$ to the most negative potential that is applied through the diodes. Since the input signals are 30 volts peak-to-peak in this case (with the negative excursion going to 0 volts) the capacitor plate that faces the diodes charges to this potential. The 220-kilohm resistor in parallel with the capacitor is the only path available for discharge, and the time constant is so long that relatively little potential is lost during each succeeding cycle of the switching frequency which, in this case, is the 3.58-megahertz locally generated subcarrier. In other words, the diodes charge the capacitor to approximately zero potential and, since the capacitor leakage rate is minimal, the capacitor voltage keeps the diodes reverse biased except on the extreme tips of the signal. When these maximum amplitude points occur, they turn the diodes on and "clamp" the output line at point $S$ to the zero potential once again.

When a signal that is in phase with the subcarrier at $A$ is applied at the input, the diodes turn on at the point where the input signal is most negative. Consequently, this point of the input chrominance information

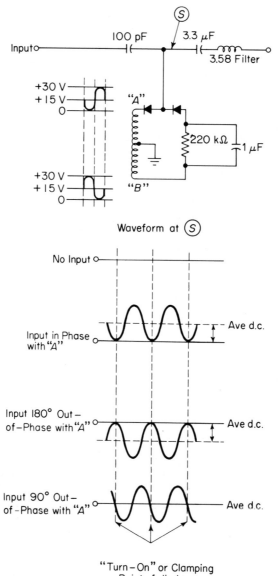

FIGURE 14-16. Clamping action of diodes switch converts input phase changes to amplitude excursions.

is clamped at zero volts and the remainder of the signal is then positive going with respect to this reference. This is illustrated in Figure 14-16.

When the input signal is 180 degrees out-of-phase with the signal at *A,* the diodes turn on when the input is at its most positive point, and this portion of the waveform is clamped at the zero potential. Thus, the resultant signal is negative with respect to this reference point and the resultant average dc potential also goes negative.

The illustration also shows an input that is 90 degrees out-of-phase with the signal at *A.* Since the diodes are momentarily switched on when this signal is passing through its zero reference point, the average dc output is also at zero. This leaves point *S* with an average that is the same as was present when there was no signal applied to the input. That is, the input signal has been clamped to the zero reference at a time when it was crossing through its own zero reference, yielding a resultant whose average is also zero.

The diagram of Figure 14-16 illustrates only one portion of the demodulation circuitry. Figure 14-17 shows the entire demodulator circuit.

The composite video waveform from the video detector is applied through 4C3 and 4L1. These two components provide a bandpass characteristic that extends approximately 1 megahertz on each side of the 3.58-megahertz subcarrier frequency. The two potentiometers at the input of 4Q6 provide for a specific amount of chrominance input to the transistor stage, and a "preset" or "manual" position for optimum ease of operation.

Emitter follower 4Q6 drives amplifier 4Q1 which has a bandpass filter in its collector circuit that narrows the bandwidth to approximately 600 kilohertz on either side of the 3.58 megahertz, with sharp cutoff characteristics at these points to eliminate luminance components from the signal.

Transistors 4Q2 and 4Q3 form a push-pull output emitter follower that provides an extremely low output impedance. The signal is then divided into two branches, one of which goes directly to the *R-Y* demodulator diodes, and the other which passes through a 90-degree delay line to the *B-Y* section. Capacitor 4C21 provides an adjustment for trimming the delay line for a precise delay time. The demodulated *R-Y* and *B-Y* signals are then routed through emitter followers 4Q4 and 4Q5 for purposes of buffering and impedance matching and applied to the output through the filter networks, 4L6, 4C18, and 4L7, 4C19, which remove any remaining traces of the 3.58-megahertz frequency component.

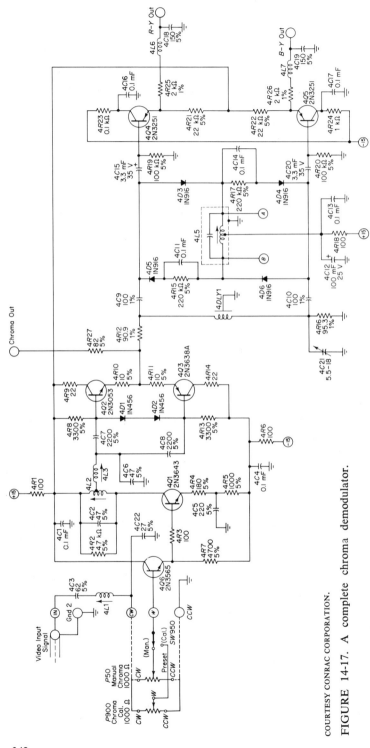

FIGURE 14-17. A complete chroma demodulator.

## MATRIXING

Once the *R-Y* and *B-Y* signals have been developed, there remains the matter of deriving the *G-Y* signal. This is accomplished by a matrix arrangement which adds specific amounts of the *R-Y* and *B-Y* signals. It has been shown mathematically that:

$$R\text{-}Y = 0.70R - 0.59G - 0.11B$$
$$B\text{-}Y = -0.30R - 0.59G + 0.89B$$
$$G\text{-}Y = -0.30R + 0.41G - 0.11B$$

Using the above equations and solving for *G-Y* in terms of *R-Y* and *B-Y* yields the following result:

$$G\text{-}Y = -0.51(R\text{-}Y) - 0.19(B\text{-}Y)$$

This states that to obtain a proper *G-Y* signal, a means must be provided whereby it is possible to add 51 per cent of an inverted *R-Y* signal to 10 per cent of an inverted *B-Y* signal. Another method would be to add these same amounts of positive *R-Y* and *B-Y* signals and then invert the resultant to obtain the proper *G-Y* signal.

It is also necessary, once the *R-Y*, *B-Y*, and *G-Y* signals have been developed, to add the luminance, or *Y*, component to each of these signals. The addition of a +*Y* component cancels the −*Y* in the above equations and the *R*, *G*, and *B* signals that were produced by the color camera are reconstructed. These three signals are then amplified and processed to drive the three guns of the tricolor picture tube.

Figure 14-18 illustrates a block diagram of the matrix action necessary to derive the *R-G-B* output signals. *R*1 and *R*2 serve to add the *R-Y* and *B-Y* signals in the proper proportion to form a −(*G-Y*) signal. The *G-Y* amplifier inverts this waveform to produce the *G-Y* signal. The *R-Y*, *G-Y*, and *B-Y* signals are then routed to the *R*, *G*, and *B* amplifiers at whose inputs the *Y* component of the video signal is added, once again by the use of matrix networks composed of two summing resistors.

Figure 14-19 is a schematic of the block diagram just examined. Transistors 5*Q*1 and 5*Q*4 act as input buffers, with 5*Q*4 also being a part of a feedback pair—5*Q*4, 5*Q*5. The feedback pair provides an accurate 1.78 gain factor for the *B-Y* signal. This assures the proper amplitude relationship between the *R-Y* and *B-Y* for matrixing and processing.

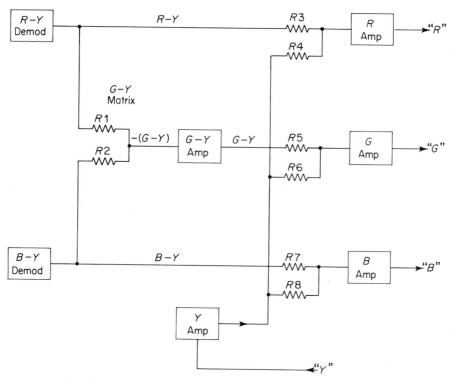

FIGURE 14-18. *R, G,* and *B* are obtained by adding luminance to the *R-Y,* *G-Y,* and *B-Y* signals.

5*Q*2 and 5*Q*6 act as transistor switches in series with the *R-Y* and *B-Y* signal path. The collectors of these two transistors also have transistor switches 5*Q*17 and 5*Q*18 connected to them in a shunt arrangement to ground through diode 5*D*3. Thus, if a positive potential is applied at input CK (color killer), the series transistors will conduct, allowing signal to pass, and the shunt transistors will be cut off (because of their p-n-p configuration) and signal will be applied to the base of 5*Q*3 and 5*Q*7.

When a negative signal is applied to the input terminal CK, the *R-Y* and *B-Y* information will not be allowed to pass. The negative potential will cause 5*Q*2 and 5*Q*6 to cease conduction and, as an additional precaution, will shunt the collector terminals to ground by causing 5*Q*17 and 5*Q*18 to conduct through diode 5*D*3.

The color killer input comes from a stage that determines the presence or absence of color information on the incoming video and, by the action of the above stages, provides a means of turning the color amplifiers off

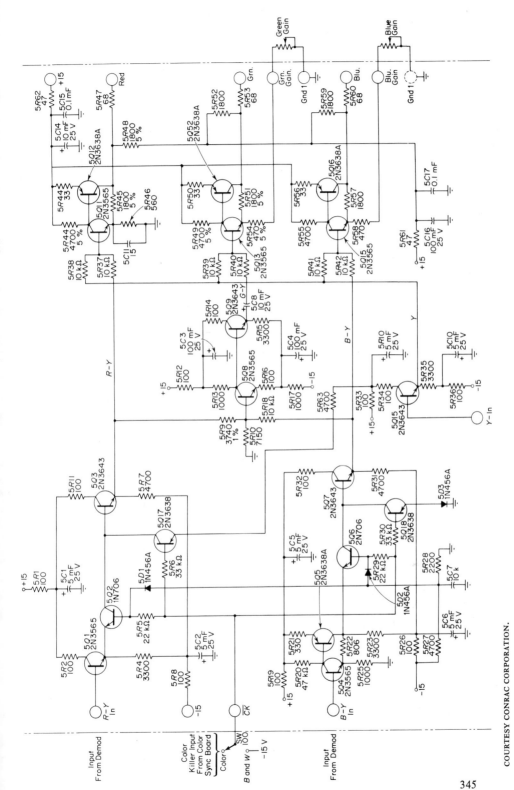

COURTESY CONRAC CORPORATION.

FIGURE 14-19. The circuit equivalent to Figure 14-18.

to assure that only the luminance information will reach the picture tube.

The *R-Y* and *B-Y* outputs from 5*Q*3 and 5*Q*7 are directed to the output amplifiers where they are individually summed with the *Y* signal that comes from the emitter of 5*Q*10. Thus, the output of these stages is the *R* and *B* drives that will ultimately excite the color picture tube.

Specific proportions of the *R-Y* and *B-Y* outputs are also routed to the base of 5*Q*8 where they are summed to form the $-(G\text{-}Y)$ signal. 5*Q*8 acts as a unity gain amplifier and inverts the signal, forming the *G-Y* signal. This output from the emitter of 5*Q*9 is felt at the base of 5*Q*13 where it is also combined with the *Y* signal to obtain the *G* signal.

Gain controls are provided for the green and blue output amplifiers, providing a means of critically aligning the output signals for best color rendition of the picture tube.

Figure 14-20 illustrates a method whereby the red, blue, and green color signals are derived independently from the demodulator diodes. The configuration here is somewhat different, with the chroma signal being applied to all three channels, and the luminance signal being added before the actual demodulation takes place. However, the end result is the same. The 3.58-megahertz subcarrier is applied to each of the demodulators with a different phase characteristic. The green channel receives the input without any alterations in phase; however, the blue channel has a delay circuit consisting of *L*1, *R*6, and *C*20, and the red channel subcarrier is delayed by *C*3, *R*3, and *L*2. These variations in subcarrier are the secret to correct color signal amplitude after demodulation.

Because the luminance signal is also applied as a part of the signal input, an *R-Y, B-Y,* or *G-Y* signal is never actually developed in this case. The luminance is already a part of the demodulated signal.

A dc signal applied to the emitters of all three stages through *R*29, *R*17, and *R*10 provides control over the brightness of the final presentation. Contrast is controlled in this case by varying the amount of luminance that is applied at the demodulators.

After the red, green, and blue video signals leave the three video drivers of Figure 14-20, they enter the final output stages shown in Figure 14-21. The outputs from these stages go directly to the cathodes of the color picture tube. These transistors are high-voltage types and the supply voltage at the input terminal is 225 volts dc. The video signals developed may exhibit amplitude variations of well over 100 volts. Such high-amplitude signals are necessary to drive the picture tube properly.

Blanking of the picture tube is achieved in Figure 14-21 by applying a positive-going blanking pulse at the emitters of the three transistors

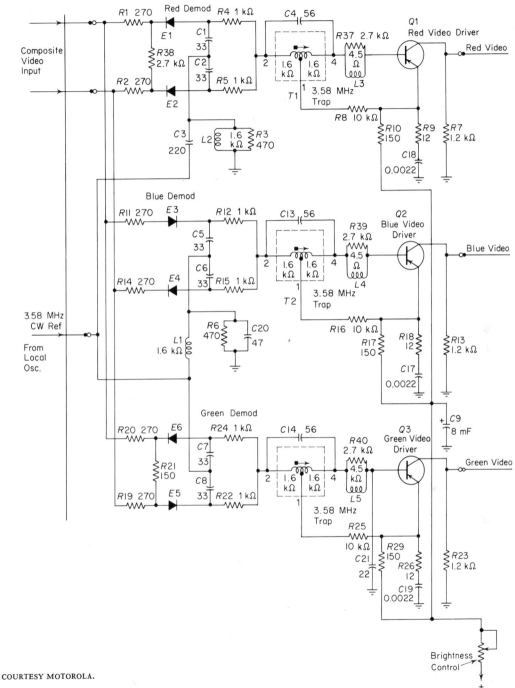

FIGURE 14-20. Demodulator converts directly to *R*, *G*, and *B*.

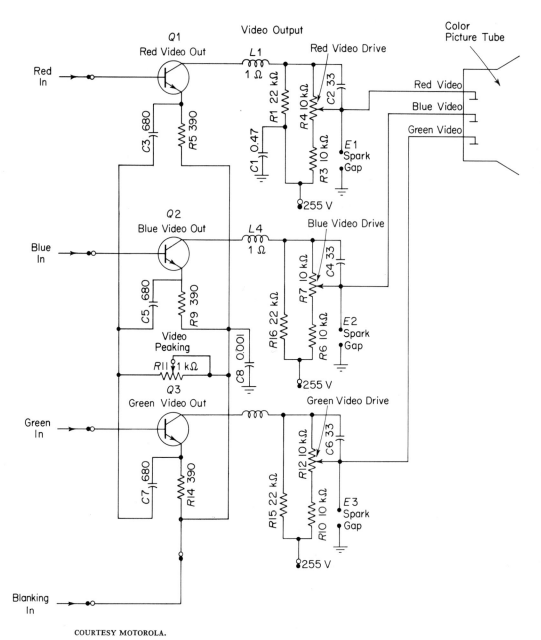

FIGURE 14-21. Color video output amplifiers.

through resistors $R14$, $R9$, and $R5$. These pulses serve to turn the transistors off during the period of horizontal retrace.

## THE SUBCARRIER OSCILLATOR

The subcarrier oscillator is a crystal-controlled oscillator whose frequency and phase must be made to match exactly that of the incoming burst signal. Therefore, some means must be made to adjust these two parameters to achieve conformance.

In Figure 14-22, the frequency of the crystal-controlled 3.58-megahertz oscillator is made variable by the action of two varactor diodes, $6D1$ and $6D2$, the capacitance of which is controlled by the dc voltage applied at the AFC input terminal. Initially, the pull-in range of the oscillator is

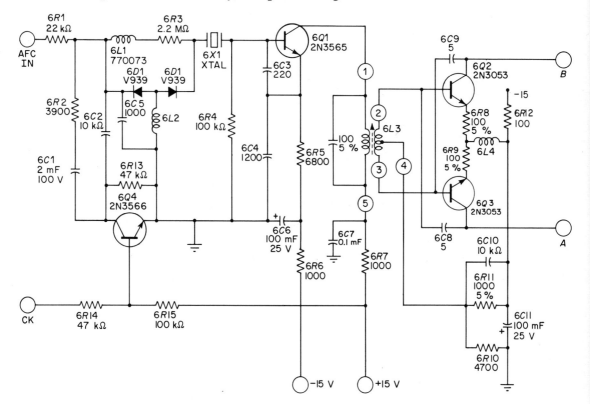

COURTESY CONRAC CORPORATION.

FIGURE 14-22. Local subcarrier oscillator in color monitor.

quite wide. This is true because the long time-constant filter circuit, consisting of $6R2$ and $6C1$, is excluded from the circuit when only a black and white signal is being received. The color killer circuit, which senses an absence of color on the incoming signal, routes $-10$ volts to the base of $6Q4$ through resistor $6R14$. With this transistor in the *off* condition, ground is removed from the network. The two components that dictate the wide pull-in range during this time are $6C2$ and $6R13$. When a color signal is being received, the oscillator will be pulled rapidly toward the correct frequency by the AFC input voltage, and a positive 12 volts will also be present at the CK input, turning transistor $6Q4$ on, effectively swamping the wide-range pull-in circuit. The circuit will then operate in a much more accurate mode and will better satisfy the accuracy requirements for phase-locking after the basic frequency has been achieved.

The output from the oscillator transistor, $6Q1$, is coupled to the output transistors through transformer $6L3$, which provides a control to vary the output amplitude of transistors $6Q2$ and $6Q3$. The output 3.58-megahertz waveforms at terminals $A$ and $B$ are equal in amplitude and opposite in phase. It is this oscillator output that would be applied to the subcarrier input terminals of Figure 14-15.

A color sync system that does not utilize a dc control voltage is shown in Figure 14-23. In this instance, the composite color video waveform is applied to the base of $Q7$, the gated color sync amplifier. However, $Q7$ is turned on only when a negative-going pulse, that occurs at the horizontal rate, is applied at the emitter. This pulse is timed to occur when the 8 or 9 cycles of burst information are present at the base, and it is this portion of the video waveform that is amplified and passed to transformer $T4S$. The 3.58-megahertz crystal converts the 8- or 9-cycle burst of gated color sync into a waveform that is continuous. The resonant action that causes this output of 3.58-megahertz information is refreshed at the horizontal rate with the arrival of each new burst interval. Thus, the output of $Q3$ is a continuous waveform of 3.58 megahertz which is then used to drive the color oscillator $Q8$. As a consequence of this method of oscillator control, the locally generated subcarrier possesses absolute phase and frequency synchronization with the color burst reference.

The oscillator output is fed to the color-oscillator phase splitter where the paraphase amplifier $Q11$ yields two signals to be summed at the input of the output amplifier by the action of the *hue* control potentiometer $R59$. The phase relationship between the two signals at this point is also governed by the length of the coaxial cable shown, and other circuit param-

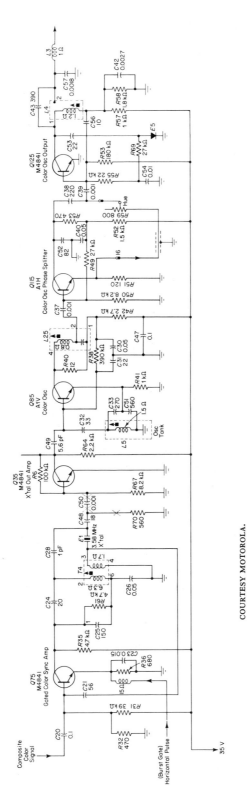

FIGURE 14-23.  Crystal amplifier duplicates burst phase.

eters. Thus, when the two signals are finally combined, they will yield a resultant product that varies in phase in relation to the setting of the *hue* control. Since the signal from the output stage, *Q*12, is routed directly to the demodulator circuitry (as shown in Figure 14-20), any variation in phase that occurs will determine the color characteristics of the television picture, hence the terminology *hue*.

## PHASE DETECTORS

For those color oscillators using a dc voltage to control the oscillator phase and frequency, it is necessary to provide circuitry that will develop the necessary voltage by comparing the incoming burst and the color oscillator waveform. Any deviation between the two should then generate a dc-voltage output, the amplitude and polarity of which will cause the oscillator to be pulled back into phase.

Figure 14-24 shows a simple phase detector that operates on somewhat the same principles as the diode demodulator in Figure 14-15. The incoming burst is applied through the transformer *T*501. The reference 3.58 megahertz is also applied to the diodes through a transformer, *T*502. The two signals are processed in such a manner that they reach the phase detector 90 degrees out-of-phase. If this relationship is maintained, there

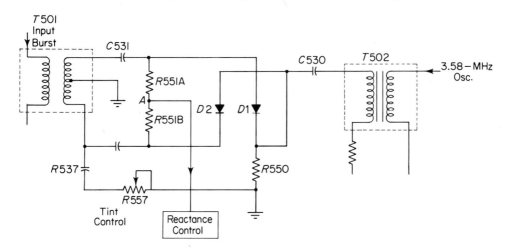

COURTESY WESTINGHOUSE.

FIGURE 14-24. Phase comparison between burst and locally generated 3.58 MHz develops an error signal to control local oscillator.

will be no output from point *A,* the junction of resistors *R*551A and *R*551B.

Assume that the phase relationship between the burst and the oscillator frequency is proper, i.e., equal in amplitude and 90 degrees out-of-phase. The applied ac voltage across both of the diodes will be equal. The ac voltage is rectified by the diodes and the resultant voltages that appear across the resistors *R*551A and *R*551B are equal. This results in zero volts of correction potential being developed at point *A*.

If the color oscillator attempts to go higher in frequency, the 90-degree phase relationship is upset and a larger ac voltage is impressed across diode *D*1. The resultant heavier conduction of *D*1 causes larger current flow through *R*551A. This same shift in frequency causes less conduction in *D*2 and a smaller current flow through *R*551B. Thus, the potential at point *A* shifts its position to a point where the proper amount of positive voltage will be routed to the oscillator to correct the phase deviation.

If the color oscillator attempts to go lower in frequency, the phase relationship is again upset, but this time *D*2 conducts heavier and the conduction of *D*1 is reduced. This causes a voltage to be developed at *A* that is opposite to that developed previously, and the oscillator receives a correction potential that tends to speed it up and again establish the proper phase with the burst signal.

The *tint* control shown in this circuit performs the same function as the *hue* control did in Figure 14-23. That is, it will vary the oscillator frequency that is ultimately applied to the color demodulators. By changing the position of the tint control, a certain amount of unbalance can be effected in the phase detector, causing a change in the control voltage that is sent to the oscillator. This, in turn, shifts the phase of the oscillator to change the color rendition of the final presentation.

## THE COLOR KILLER

In color receivers and monitors, it is the purpose of the color killer circuitry to disable the chrominance circuits when there is a black and white signal being received. This assures that only luminance information will make its way to the picture tube.

Referring again to the crystal output amplifier in Figure 14-23, it can be seen that there will be no 3.58 megahertz present at the collector of this stage when burst is not present at the input. Therefore, if this point is connected to the input of the circuitry in Figure 14-25, any 3.58 megahertz that is present will be rectified by diode *E*2, filtered, and applied to the base of *Q*4, the *automatic color control* amplifier. *Q*4 acts as a switch, turning on when the burst-generated 3.58 megahertz is present at the input. As a result, the switching action of this stage can be used to control the color processing circuits, turning them on when a color signal is present at the input and turning them off when black and white only is being received.

Many variations of this circuit are used, but all have one thing in common; they sample the incoming video waveform for burst presence. When no burst is present, the chrominance circuits are disabled and a monochrome presentation is achieved.

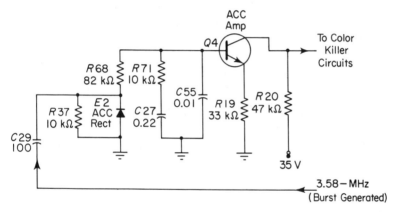

FIGURE 14-25. Automatic color control amplifier turns chroma circuits off when no burst is present on the input.

## THE COLOR PICTURE TUBE

The color picture tube widely used in television receivers throughout the United States is the three-gun, shadow-masked tricolor kinescope. Color is generated by exciting minute phosphor dots that cover the face of the tube with three beams of electrons.

The dots that cover the face of the tube are arranged in trios of red, green, and blue, with each individual color being energized by its respective electron beam. There are hundreds of thousands of these dot trios on the faceplate of modern color picture tubes. Because of the small size of the phosphor dots, the human eye does not discern each as a specific element of red, green, or blue light, but integrates them into areas of colored light, with the resultant color being determined by the relative intensities of the three primary colors being emitted by the dots.

### The Electron Gun Assembly

The three high-velocity electron beams are generated by electron gun assemblies positioned in a triangular configuration 120 degrees apart at the extreme rear of the tube. All are tilted slightly toward the centerline of the tube to cause all three beams to converge near the faceplate.

Each of the electron guns in the color tube is basically the same as the single gun used in monochrome picture tubes. The function of each gun is also identical to its monochrome counterpart. That is, it must generate, direct, focus, and modulate an electron beam. However, in the color picture tube each beam exists to illuminate only one set of color phosphor dots. Figure 14-26 shows an exploded diagram of the gun structure in a typical color tube.

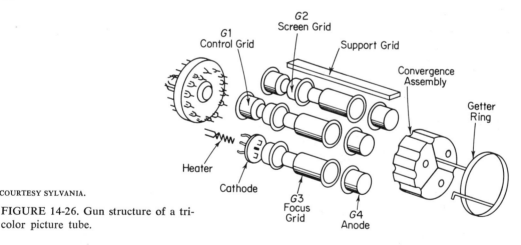

COURTESY SYLVANIA.

FIGURE 14-26. Gun structure of a tri-color picture tube.

The heater and cathode, of course, perform the normal tasks of releasing electrons to form the beam. Depending upon the particular circuit configuration of a receiver, either the cathodes or the $G1$ elements may be used to modulate the beam. Thus, proper control of the voltages applied to either of these two elements creates the changes of hue that constitute the color picture. The $G2$ screen grid is generally operated at about 700 volts and acts to attract the electrons emitted by the cathode, producing a directional flow of electrons to form the beam.

The *G*3 focus grid interacts with the screen grid and anode to create two fields that act as electrical lenses to focus the beam to the smallest possible dot on the phosphor screen. The *G*3 element is usually operated at a potential of several thousand volts.

The *G*4 anode element is operated at extremely high potential, approximately 25,000 volts, to accelerate the beams to the high velocity necessary to excite the phosphor dots.

Located at the front of the gun assembly structure is the *convergence assembly:* a magnetic shield which allows each beam to be magnetically shifted in position without appreciably affecting the others. A *convergence yoke* (not the deflection yoke) is positioned on the tube exterior to place the red, green, and blue convergence coils directly above their respective beams. This is shown in Figure 14-27. The arrows used in conjunction with the beam elements illustrate the effects of the magnetic field upon each beam.

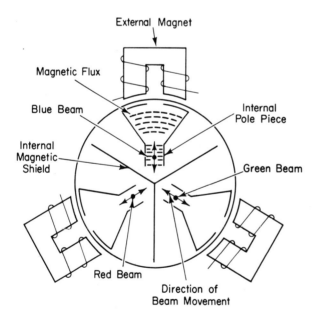

COURTESY SYLVANIA.

FIGURE 14-27. Convergence coils control position of each individual beam.

### The Aperture Mask

Directly behind the phosphor dot screen is a thin metal sheet perforated with small holes approximately 0.01 inch in diameter. There is one hole for each dot trio on the tube faceplate. Thus, for the electron beams to reach their respective phosphor dots, all three must converge at this *shadow mask,* or *aperture mask,* and pass through the small opening (as illustrated in Figure 14-28). Proper convergence of the three beams and correct placement of the mask thus assures that each beam will strike only one particular color of phosphor dot.

Figure 14-29 is a cutaway drawing of the color picture tube. The magnified area shown in circles at the faceplate shows the placement and action of the aperture mask with respect to the phosphor screen. The three beams are scanned across the tube face by means of an external electro-magnetic deflection yoke. The converged beams therefore move successively from one opening in the mask to another, passing through the opening and illuminating the phosphor momentarily before moving on to the next opening. This means that the beams are not striking the phosphor faceplate at all times, but are actually blanked out while moving between openings.

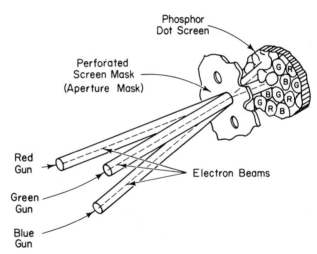

FIGURE 14-28. Red, green, and blue beams converge at the aperture mask.

For this reason, the brightness capabilities of the color picture tube are somewhat limited. This was especially true in early color sets, but modern phosphors and improved methods of applying them, coupled with the high anode and accelerating voltages used in color tubes, have greatly improved the situation.

It is quite apparent that the placement of the aperture openings in the shadow mask must be precisely positioned with respect to the dot trios. To assure that this requirement is met, the same shadow mask is often used photographically to place the dots on the faceplate and then is actually used in the construction of the tube. Thus, the shadow mask is matched to the faceplate, and any irregularities in the hole structure of the shadow mask will also be incorporated in the phosphor dot arrangement and no non-linearity of convergence will occur.

Because the electron beams are continuously bombarding the shadow mask when moving between aperture openings, the mask dissipates several watts of power. The mask naturally expands as the heat builds up, but modern designs assure that expansion movement is in line with the electron beams. This movement will, therefore, not affect the convergence of the beams or the purity of the color.

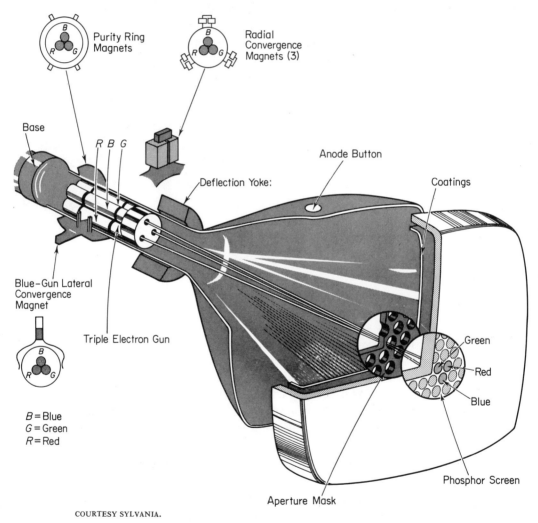

FIGURE 14-29. Exploded view of three-gun tricolor picture tube.

## Static Convergence

It has been stated that the three electron guns are tilted slightly to achieve convergence. However, in practice it is not possible to position them accurately enough to achieve acceptable results. For this reason, a series of adjustable magnets are positioned around the neck of the tube. It is the purpose of these magnets to position the individual electron beams properly for convergence at the center portion of the phosphor screen. There is one magnet for each beam and they are logically called the red, green, and blue magnets. An additional magnet is provided for the blue beam to allow it to be moved horizontally as well as vertically, as illustrated in Figure 14-30. With the beams positioned near the center of the tube, adjusting these magnets will achieve *static* convergence.

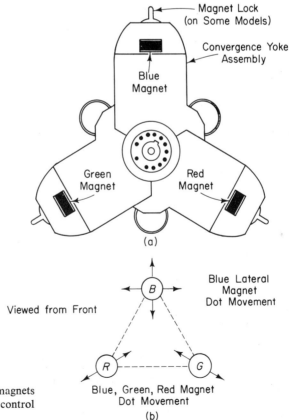

COURTESY WESTINGHOUSE.

FIGURE 14-30. Permanent magnets positioned around the tube neck control purity adjustments.

### Dynamic Convergence

When the beams are being scanned across the phosphor screen, a situation occurs that cannot be corrected by the use of simple magnets. Figure 14-31 shows how the actual point of convergence will change with respect to the face of the tube. The beams are being scanned in a method that will cause the actual point of convergence to traverse an arc. If the picture tube was made so that the faceplate conformed to this arc, the problem would not be especially difficult. However, most modern picture tubes have flat, rectangular screens and a method known as *dynamic* convergence must be used to achieve convergence over the total screen area.

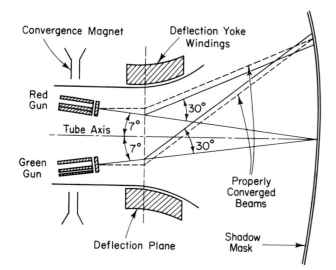

COURTESY WESTINGHOUSE.

FIGURE 14-31. Effects of electron beam scanning in multigun picture tube.

Dynamic convergence techniques also exert magnetic influence over the position of the red, green, and blue electron beams. It differs from static convergence in that convergence coils are used and their magnetic fields are controlled by parabolic waveforms that are generated in conjunction with the deflection waveform.

The problem of dynamic convergence may be better understood by referring to Figure 14-32. In this example, only the red and green guns are shown. With the guns angled as shown, the two beams will converge at the tube faceplate in a proper manner. However, when the beams are deflected for scanning, both are deflected by exactly the same angle $\theta$. The

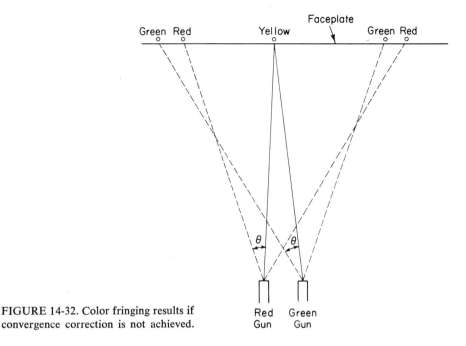

FIGURE 14-32. Color fringing results if convergence correction is not achieved.

result is a convergence point that lies to the rear of the phosphor dot screen by a distance determined by the total angle of deflection.

Of course, in actual application there are three electron guns to consider, and convergence problems therefore occur both horizontally and vertically, so correction waveforms must be generated at both a horizontal and vertical rate. Each of these waveforms (Figure 14-33) may be considered as simply one cycle in a series which occurs continuously during active scanning. The individual red, green, and blue waveforms are applied to respective convergence yoke coils which are positioned above each gun.

The physical location of the three guns, 120 degrees apart and approximately parallel to the centerline of the tube, demands that each waveform be slightly different. The blue gun located at the top of the three-gun pyramid is the only one that can be thought of as being centered horizontally, as it lies in the center plane of the tube. (None of the guns are vertically centered.) Therefore, the horizontal waveforms show only the blue waveform to be symmetrical, or not "tilted," to overcome the physical offset from center. The other waveforms are modified by a tilt factor which is a function of the amount of beam separation that must be overcome.

By way of example, consider the blue beam which is situated above the tube axis. For proper convergence to occur during vertical deflection, a vertical correction waveform (as shown) must tilt the beam toward the screen top. However, while scanning the bottom of the screen, it travels further and a stronger convergence field must be used. The reverse is true for the red and green beams that lie on the opposite side of the tube axis.

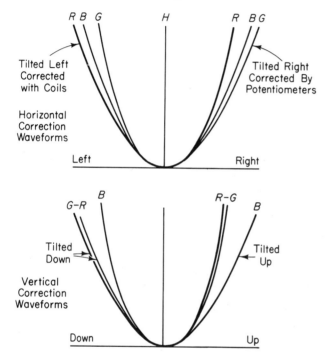

FIGURE 14-33. Dynamic convergence waveforms.

It is interesting to note with reference to the horizontal parabolic waveforms that the blue correction waveform is not tilted because the blue beam is actually projected along the vertical axis of the tube. Because the red and green electron guns are located on opposite sides of this axis, their respective waveforms are tilted opposite to one another.

### Vertical Convergence Circuitry

Figure 14-34 illustrates circuitry used to generate vertical convergence waveforms. The input to the base of $Q1$ is the vertical sawtooth deflection signal. It is processed into a parabolic waveform which is present at terminal 1 of $T1$ and applied at a 60-hertz rate to the red, green, and blue vertical convergence coils. The amount of convergence current that will be felt through the coils is determined by the settings of $R18$ for the red and green, and $R19$ for the blue coil.

The tilt of the parabola is incorporated by the addition of an adjustable sawtooth current supplied by the *vertical tilt* windings 6, 7, and 8 of transformer $T1$. The sawtooth waveforms at 6 and 8 are opposite in polarity and the *RG vertical tilt* and *blue vertical tilt* controls, $R15$ and $R16$, determine the amplitude and polarity of the sawtooth waveform that is added to the parabola, thus regulating the tilt of the final waveform.

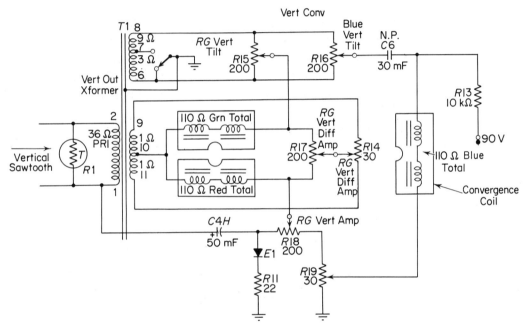

FIGURE 14-34. Circuit develops vertical convergence waveforms.

It is not desirable to provide identical waveforms for the red and green coils because the electron-optical distortion for these two guns is in opposition. That is, as the vertical deflection angle increases, the red and green beams tend to need waveforms that are tilted opposite to one another. Therefore, *differential amplitude* and *differential tilt* controls $R17$ and $R14$, are provided for adjustment to the required differences between red and green coils. $R14$, the *differential tilt* control, accepts a positive pulse at one end from terminal 9 of the transformer. A negative pulse is applied to the other end of the potentiometer from terminal 11. Depending upon its position, the arm of $R14$ delivers a positive, negative, or zero amplitude pulse to $R25$. The differential amplitude potentiometer, $R25$, determines which coil, red or green, receives the largest amount of the selected correction signal from $R14$. This procedure forms a mirror image correction signal for the red and green electron beam correction coils. Because the blue coil lies on the vertical axis, it does not need any differential correction.

### Horizontal Convergence Circuitry

In Figure 14-35 the horizontal convergence waveforms are developed by use of a driving pulse derived from the flyback transformer.

A horizontal parabolic waveform for the blue convergence coil is developed by two $L/C$ circuits. $C7$ and $L2$ in parallel with $C8$ and $L3$ create the convergence signal from a positive-going flyback sample which is impressed across potentiometer $R24$. $R24$ provides an adjustment of the horizontal parabola amplitude, and variable inductors $L2$ and $L3$ set the tilt and center correction for the blue gun.

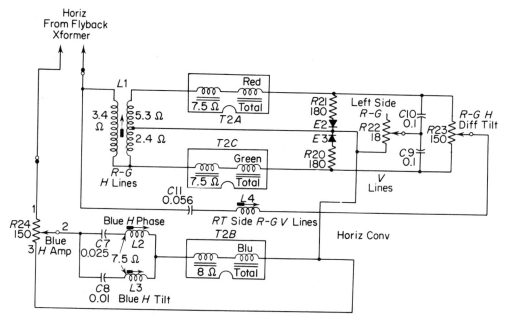

COURTESY MOTOROLA.

FIGURE 14-35. Horizontal convergence waveforms are initiated from the horizontal flyback pulse.

The red-green convergence signal is developed by applying a positive pulse of about 250 volts amplitude, occurring at the horizontal rate, to transformer $L1$. The center-tapped secondary of $L1$ provides two signals 180 degrees out-of-phase across $R23$, a potentiometer that controls the red-green differential tilt of the parabolic waveform. Depending upon where the potentiometer is set, the wiper arm selects a positive or negative parabolic waveform to be routed to $L4$ which, in turn, affects the primary signal at transformer $L1$. The resulting waveform modification determines whether the red or green convergence coil receives the larger amount of correction. Potentiometer $R22$ adjusts the amplitudes of current through both coils equally. Such an arrangement controls amplitude and direction of tilt of the correction signals.

## CONVERGENCE PROCEDURES

Convergence procedures will vary somewhat with receivers and monitors made by different manufacturers, however, all are governed by the same general principles and concepts. The following discussion should, therefore, be considered as simply an example of one of the many methods that may be employed.

### Degaussing

Before any attempt is made at converging a color monitor or receiver, it is generally a good idea to completely degauss the unit. Many modern receivers contain an automatic degaussing device within the cabinet, but this unit is included only to achieve elimination of relatively low levels of acquired magnetism. Without exception, a manual degaussing coil is preferred for use on a receiver or monitor during initial adjustment.

### Purity Adjustment

Purity adjustments must be made while observing a single color on the face of the tube. Generally red is the preferred color with which to begin the adjustments. This means that the blue and green guns must be disabled by turning off the screens or providing a means of turning the guns off by applying reverse bias to the control grids. Once the red field is established it should be examined to ascertain that no areas of impurity are present. If there are areas where the color deviates from the desired red, it is usually possible to correct the problem by adjustment of the purity ring magnets located on the neck of the tube. These should be adjusted for the best overall red display. Once the red is adjusted, the green and blue fields may be checked for similar results. If the red has been properly adjusted, the blue and green should require no correction.

### Static Convergence Adjustments

For best results in converging a color set it is desirable to utilize a crosshatch or dot-bar generator to develop the sharp-edged lines that are necessary to observe for proper convergence. Static convergence is generally performed using the dot pattern as a reference and adjusting the three permanent magnets located on the convergence assembly. These three magnets, coupled with the blue lateral magnet that is separately mounted on the tube, are adjusted to achieve both horizontal and vertical static convergence.

If the blue gun is cut off, the red and green convergence magnets may then be adjusted to achieve a yellow dot at the center of the screen. When the red and green dots are exactly superimposed there should be no visible fringing of the red or green, only a yellow dot. With the blue gun reactivated, it should then be adjusted with the blue magnet and blue lateral magnet to achieve coincidence with the yellow dot. With the three converged, the resultant dot should appear white, with no obvious fringing.

Once again, static convergence is performed to achieve convergence on only an extremely small portion of the center area of the picture tube screen. Other areas of the screen should be ignored during this procedure.

### Dynamic Convergence Adjustments

The dynamic convergence controls provided on color monitors and receivers are for the purpose of obtaining correct convergence at the edges of the picture. In this procedure the bar pattern is generally selected from the crosshatch generator to provide a continuing sharp image along the edges of the screen to facilitate convergence procedures.

By way of example, a dynamic convergence procedure for a Westinghouse color receiver follows. The convergence panel layout shown in Figure 14-36 may be considered as typical of those provided on many commercial receivers. In addition, those displayed bars from the crosshatch

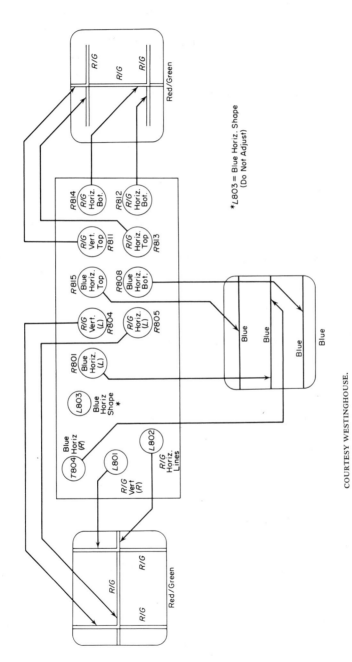

FIGURE 14-36. A typical convergence control panel.

generator that are affected by the individual controls are shown with identifying arrows. The adjustment procedure is as follows:

1. Bias *off* the blue gun (this may be done by applying a negative voltage of approximately 100 volts to the control grid through a resistor of approximately 100 kilohms).
2. Adjust $R812$ for bottom center $R/G$ lines.
3. Adjust $R813$ for top center $R/G$ horizontal lines.
4. Adjust $R814$ for bottom $R/G$ center vertical lines.
5. Adjust $R811$ for top $R/G$ center vertical lines.

(When properly adjusted, the bars should appear yellow with minimum color fringing of red or green apparent)

6. If necessary, readjust the static convergence at this point, and then repeat the above steps, 1 through 5.
7. Reactivate the blue gun by removing the negative bias. Apply the negative bias to the green control grid to turn off the green gun.
8. Adjust $R801$ for left blue horizontal center lines.
9. Adjust $T804$ for right blue horizontal center lines.
10. Adjust $R808$ for bottom blue horizontal lines.
11. Adjust $R815$ for top blue horizontal lines.
12. If necessary, readjust static convergence and repeat steps 7 through 11.
13. Remove the negative bias from the green gun and again apply it to the blue gun.
14. Adjust $L802$ for right $R/G$ horizontal center lines.
15. Adjust $R805$ for left $R/G$ horizontal center lines.
16. Adjust $L801$ for right $R/G$ vertical lines.
17. Adjust $R804$ for left $R/G$ vertical lines.
18. Remove the negative bias from the blue control grid.

This completes the convergence procedures for this receiver and should result in a color picture that is properly registered and exhibits little or no color fringing during normal operation.

It should be noted that perfect convergence can never be accomplished because of the many variables and distortions that can influence the results. However, excellent convergence can usually be effected by patient performance of the manufacturer's recommended techniques.

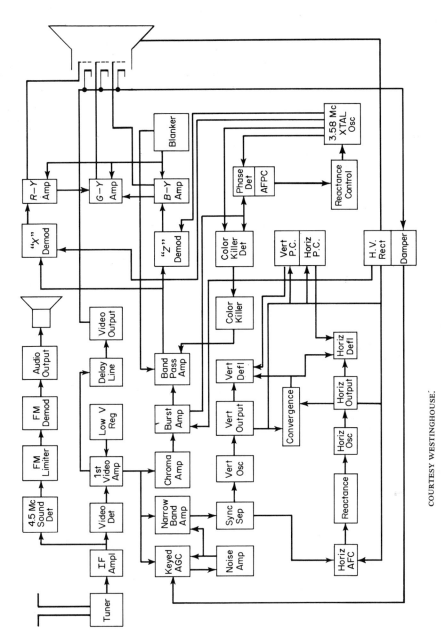

FIGURE 14-37. Block diagram of complete color receiver.

## THE COMPLETE COLOR RECEIVER

Figure 14-37 is a block diagram of a color receiver. It shows the method in which the various circuits that have been discussed may be interconnected in a typical application. As previously indicated, many of the blocks shown have almost identical counterparts in a monochrome receiver, but parameters such as flat bandpass characteristics, signal delay times, and circuit stability weigh much more heavily in a color receiver. The problems encountered in the setup, operation, and maintenance of color receivers and monitors are also more complex. However, the millions of color sets now in use testify to their reliability and practicality.

# 15

# VIDEO SWITCHING SYSTEMS

In video systems that employ more than one camera, it is often necessary to have the capability to present the various individual video signals on a single monitor at specific times, and to be able to route the video signals to several other points by some form of switching network.

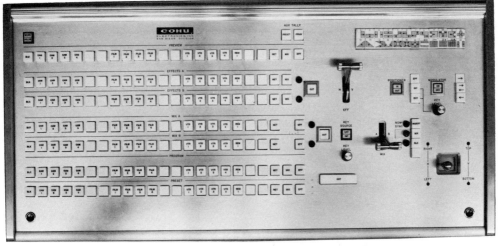

COURTESY COHU ELECTRONICS, INC.

A broadcast video switcher control panel.

Use of a video switcher system makes it possible to program the display on one or a number of monitors by sequentially switching from one video signal to another. The results obtained from these switching procedures are quite familiar to anyone who has watched commercial television

371

programming. Through switching, the rather restricted two-dimensional image presented by one camera system can be given additional perspective by changing the display to that of a second camera system that views the same scene from a somewhat different angle. Also, information being viewed by a number of cameras at various locations can be presented on a single monitor by a switching system that has provisions to accommodate as many video inputs as are required. This gives the monitor a much broader ability in an application of information display.

COURTESY DYNAIR ELECTRONICS, INC.

A video switcher for educational and industrial television.

The ultimate destination of the outputs from a video switcher may be a transmitter, video tape recorder, etc., as well as a monitor; and all such "using destinations" have specific requirements as to the video signal that is being sent to them. Some of the more stringent requirements demanded by particular applications can be met only by sophisticated switching systems.

### Methods of Switching

A video switch is a device that is used to turn a video current on or off. As such, its configuration might be as simple as that shown in Figure 15-1. In this illustration a common two-position mechanical switch is used to select which of two camera systems will be viewed on a monitor. While this basic method may be perfectly acceptable in some applications, its limitations would be quite severe in situations such as that shown in Figure 15-2.

Figure 15-2 illustrates an example of video switching that allows each of five remotely located viewing sites to select any of three camera outputs independent of the other viewing sites. Such a system is termed a $3 \times 5$ switching system, deriving its definition from the three inputs and five outputs. If four cameras and ten monitors were involved, it would be a $4 \times 10$ system.

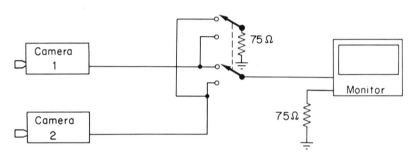

FIGURE 15-1. A simple video switch.

If mechanical switches located at each of the viewing sites were the only means used to switch the video in Figure 15-2, the problems involved would become quite pronounced. If the viewing sites were at widely separated points, for example, the cable from each camera system would have to be routed to each one, and the cable losses at the video frequencies involved could render the system unusable or demand the use of several distribution amplifiers. Also, it is generally necessary to provide isolated outputs to each monitor so that switching at one point does not visibly disturb the images at the other locations. Mechanical switching does not readily lend itself to an easy answer to such problems.

Electromechanical switching, involving the use of centrally located relays which are energized by remote switches, solves many of the problems encountered in purely mechanical methods. Probably the greatest advantage lies in that it does away with the great bulk of cable needed for signal distribution in a mechanical system.

When electromechanical switching is used, pushbuttons on control boxes located at various viewing sites would likely energize a relay within a matrix at a central point and allow the video transfer to be effected. Thus, there is no need to route camera signals to any destination other than the central matrix. Properly located isolation amplifiers could assure switching at one viewing site without interference being observed at the other monitors.

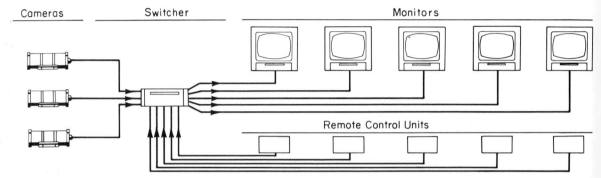

FIGURE 15-2. Remote control units activate centrally located switcher matrix.

Present day electromechanical switching devices may typically use dry-reed relays. These devices have solved many of the problems encountered in previous relay switching equipment. Switching times are much faster and, when precisely timed, transitions may be made during the vertical blanking interval, as is required in many applications. Electromechanical devices may still have shortcomings in the areas of long-term reliability, contact bounce, inductively induced transients, and relative speed.

The electronic switch, and specifically the solid-state video switch, overcomes many of the problems encountered in the mechanical and electromechanical systems. Switching speed is much higher, with transition times of much less than a microsecond being common. Size is generally much smaller and, due to the inherent reliability of solid-state devices, maintenance is reduced to a minimum.

### Switcher Classification

Video switching systems generally may be grouped into two major categories: distribution-type switchers, and broadcast- or program-type switchers. Figure 15-3 shows a matrix diagram of the 3 × 5 system discussed previously. This is an example of the distribution-type switcher system, and it should be noted that all inputs are available to all monitors.

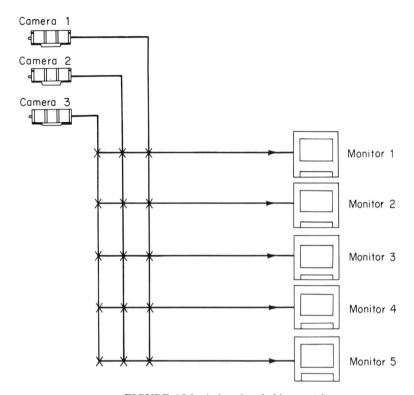

FIGURE 15-3. A 3 x 5 switching matrix.

The X's indicate the possible combinations that may be achieved. If the crosspoint indicated by the X on the intersection of the lines from camera 2 and monitor 1 were selected, the scene viewed by camera 2 would appear on monitor 1. At the same time, monitors 2 through 5 may be selected to view any camera that is desired. Isolation amplifiers are not depicted in this simplified matrix illustration, but it should be understood that they must be included to eliminate transients and level changes that would be caused by switching operations at the various stations.

The broadcast-type, or program, switcher differs from the distribution type in that the primary concern is to program various camera signals so that they will be acceptable for routing to one final destination, the transmitter. Other outputs are generally provided for such things as preview monitors, video tape recorders, and the like.

A

B

FIGURE 15-4. A large distribution switcher system.

The distribution switcher may typically be much larger in the number of active crosspoints required. A specific application may call for 20 inputs and 100 outputs. This would require the rather awesome number of 2000 pushbuttons, if a button is to be provided for each crosspoint. Also, since lighting of the pushbutton is generally the method used to determine activated crosspoints, there must be provided some 2000 switching circuits for an equal number of lamps. Switching systems that approach or exceed this size will generally be computer controlled or will incorporate a digital-type address system to activate the crosspoints and may utilize alphanumeric readout devices to display the switcher configuration.

To illustrate the possible size of a distribution switcher, Figure 15-4 shows two views of a solid-state electronic switching system of the distribution type that incorporates some 28,500 crosspoints. It was designed to be operated by an IBM 7090 computer. The equipment shown does not include control or readout devices.

## TYPES OF SWITCHING TRANSITIONS

Figure 15-5 is an illustration of several types of switching transitions that may be effected with various types of devices.

Figure 15-5A illustrates the *gap* switch, which provides transfer from one scene to another but produces a blank time interval between the two. This causes picture information to go black during the transition. *T*1 illustrates turn-off of one scene and *T*2 shows turn-on of the second scene. Such a gap in the video information may be caused by the transition time required for the arm of a switch or relay to move from one contact to another. It is generally not desired because it can cause a noticeable dropout in the picture information during switching. This could be especially noticeable between two scenes with fairly high brightness levels.

To eliminate the problems associated with gap switching, mechanical and relay switchers often incorporate make-before-break switching contacts. This gives rise to the *lap* or *overlap* switch as illustrated in Figure 15-5B. When the lap transition is used, scene number 2 is turned on just before scene number 1 is turned off. This type of transition was found to be less objectionable from a viewer standpoint than the gap method of switching. It is, of course, desirable to complete the switching in the minimum amount of time. Modern methods allow switching times to be measured in nanoseconds where earlier nonelectronic switching made transfer times of 500 to 1500 microseconds acceptable.

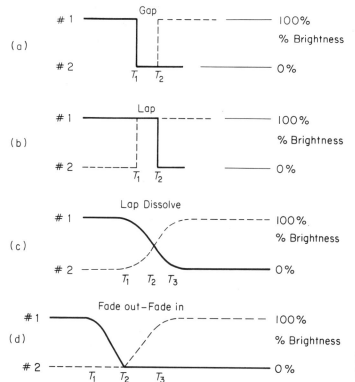

FIGURE 15-5. Four types of switching transitions.

Current broadcast and other program switchers generally incorporate some method of *vertical interval* controlled switching. This term describes the function of a switcher that transfers from one scene to another only during the vertical blanking interval. Switching in this manner eliminates any visible evidence of switching that might be observable as a disruption during the normal vertical scan. However, it is necessary that all sources of video information be in phase with each other, so that when transfer is initiated during the vertical blanking interval of one signal, it will be completed within the vertical blanking interval of the second signal. The actual switching transition may be either a *lap* or *gap* operation.

Figure 15-5C shows the signal resulting from a *lap-dissolve* form of switching. This might be accomplished by two potentiometers connected to the two signals that are to be switched. Signal amplitude is slowly reduced on the number 1 signal while the number 2 signal amplitude is increased at the same rate. This may be done directly through the resistors, or they may be used to control the gain of two amplifiers whose outputs are then combined.

Figure 15-5D illustrates the *fade out–fade in* method of video transfer. It can also be done with the two potentiometers mentioned for the lap-dissolve. In this case, by separate actuation of the two potentiometers, signal number 1 is slowly reduced in amplitude until 0 per cent signal level is obtained, and then signal number 2 is slowly brought from 0 to 100 per cent by the second potentiometer.

The examples of lap-dissolve and fade out–fade in are generally performed in a switcher by the incorporation of a lap-dissolve amplifier. This unit is also sometimes called a *mixer* or *fader*.

## SWITCHER CONFIGURATIONS

Figure 15-6 is a functional block diagram of a very simple broadcast-type switcher. It has seven inputs that may be selected to drive either of two output amplifiers. These, in turn, are feeding into a remotely controlled *mix amplifier,* or *mixer*. The mixer transfers video by the *lap-dissolve* or *fade out–fade in* method. Assume that the potentiometers at the remote mix control were positioned so that the *A* input was at 100 per cent input level and the *B* input was at 0 per cent. If crosspoint number 2 was selected on the *A* buss, it would then be appearing at the output of the mixer. Assume that crosspoint number 4 were selected on the *B* buss. If the levers that control the potentiometers were then moved their

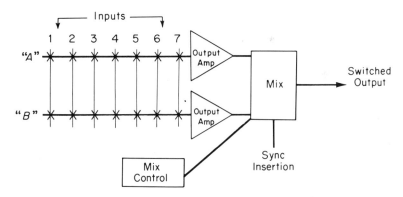

FIGURE 15-6. Simple switcher allows "mixing" busses *A* and *B*.

full travel, the output from the mix amplifier would transfer from the *A* buss to the *B* buss at a relatively slow rate and the video display would show a lap-dissolve from camera number 2 to camera number 4.

With the mixer controls positioned so that the *A* buss is selected, sequential operation of the crosspoints on the *A* buss would allow *lap, gap,* or *vertical interval* switching, depending upon the type of switching matrix in use. The fact that the mixer is shown with a sync input indicates that sync is added to the video at this point and illustrates the fact that many of the input video signals may be noncomposite in nature. In this instance it is imperative that all inputs be synchronous so that no matter which video input is selected, the sync waveform will be inserted at the proper place.

Figure 15-7 illustrates a system which, by the addition of two more output busses, provides a means for switching to the output without going through the mixer. This then frees the mixer to be programmed out on the *preview* buss. In a broadcasting situation, this would allow the video taping of a program from the *preview* buss through the mixer, while using the

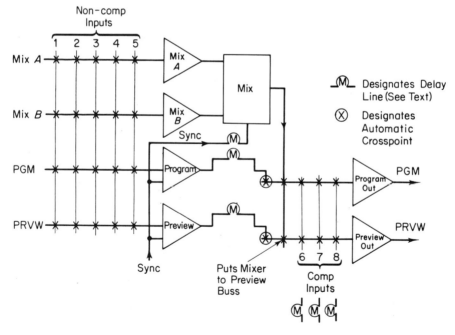

FIGURE 15-7. A somewhat more complex switcher.

*program* buss to route a different program to the transmitter. The cross-point on the preview buss that must be selected for this operation is indicated by an arrow. Directly to the left of this crosspoint is another cross-point that is enclosed within a circle. This is the schematic designation for an *automatic crosspoint*. The designation *automatic* means that it is not directly controlled by a specific pushbutton or external command. Whenever one of the noncomposite input crosspoints (numbers 1 through 5) is energized on the preview buss, the automatic crosspoint must also energize to allow the output of the preview isolation amplifier to get out on the preview output buss. In the event the mixer is selected to go out on the preview buss, the automatic crosspoint would not energize. This provides isolation of the output from the delay line shown and the capacities within the preview isolation amplifier.

At the output of the amplifiers on the program and preview busses, and inserted into the sync input which is being routed into the mixer, can be noted three delay lines with the designation $M$. These delay lines are provided to delay the signals that pass through them, by an amount equal to the time that it takes video information to pass through the *mixer,* hence the designation $M$. Information that goes through the mixer may experience from 15- to 200-nanoseconds delay while being processed by its circuits. While this may not seem excessive by some standards, it should not be forgotten that, when switching color signals, a phase shift of one or two degrees on the 3.58-megahertz color information can be objectionable. Therefore, when switching between two color signals, the phasing of the chroma information should be held to as close a tolerance as is possible.

Delay lines, such as those shown in the diagram, can take on different configurations. Often, coaxial cable is used. RG-59 coaxial cable, for instance, gives approximately 1.5-nanoseconds delay per foot. Therefore, to compensate for the delay of approximately 150 nanoseconds that was experienced in the mixer, there would have to be three 100-foot lengths of cable inserted, as shown, to assure that all possible switching configurations would provide an output that is always identical in phase.

It is important to note that whenever cable or other types of delay lines are used to provide synchronization of phasing between all signals, the cable losses that result at the various frequencies must be equalized to maintain a flat overall frequency response. This may be achieved prior to entering the output amplifiers as shown in Figure 15-7.

It may be seen that the *program* output amplifier has crosspoints at its input that receive signals from both the delay lines (which have caused high frequency loss) and the mixer. Since the mixer probably does not affect the response appreciably, the amount of equalization that this signal needs will not approach the magnitude of the equalization needed by the signal that passed through the delay line. Consequently, equalization circuitry in the output amplifier would not assure equal response of all signals. However, if each isolation amplifier preceding the delay lines were peaked in such a way as to provide compensation prior to entering the delay line, this would effect the proper results.

Figure 15-8 is a further expansion of the previous example. Two more busses have been added and we now have the additional capability

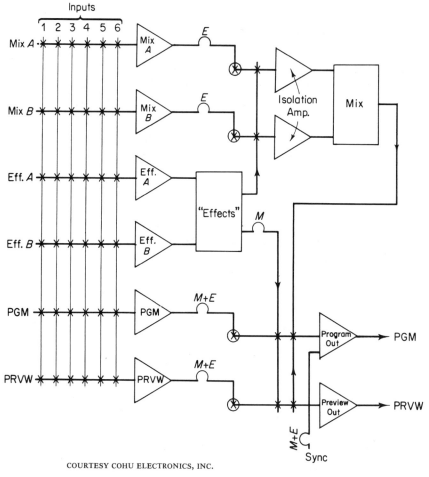

FIGURE 15-8. A "single re-entry" switching system.

of a *special effects generator,* or *effects.* An effects generator generally does one of three things:

1. Effects can "key" the *Eff B* into the *Eff A* input as shown in Figure 15-8. Assume that the *A* input is selected and is going out to the *program* line. It is possible, by the action of the effects generator, to switch to the *B* signal every time that the *B* signal exceeds a predetermined level. This means that it is possible to switch to the *B* signal several times during the course of one horizontal line of video information. This is a common practice in broadcasting and is sometimes used as a means of adding titles to a video scene.

2. The effects generator can place the *B* signal onto the output in place of the *A* signal based upon signals derived from internal signal generators.

3. The effects generator can key the *B* signal into the *A* signal based on a third signal that may be derived from an external signal generator or even a third video signal. Using the latter as an example, it is possible to obtain the keying signal from a camera that is observing a white cutout placed on a black background. Therefore, everytime the white signal from this source occurs, it will key in the *B* buss and a return to the black signal will key in the *A* buss.

The diagram that was shown in Figure 15-8 depicts a *single re-entry* switching system. In the case shown, this means that the output of the effects generator can be fed into the mixer. This allows the operator to fade into an "effects," fade away from it, lap-dissolve into an "effects," or lap-dissolve an "effects" into a previously established background. Broadcasters generally use this to good advantage, and its use can be seen routinely on commercial stations.

Another example of a single re-entry system might have the effects generator and the mixer interchanged as to position within the switcher. This would still be a single re-entry type of system. Placement is primarily dictated by the application and can be determined by the user.

It will be noted that delay lines once again play an important role in the illustration. Because of the three possible amounts of delay incorporated within the switcher, three different lengths of delay line must be used. Since the mixer may or may not be fed through the effects generator, those inputs that are not routed through the effects must be delayed by an amount equal to that experienced by signals that *do* go through the effects circuitry. Likewise, if the effects is switched into the program line but not

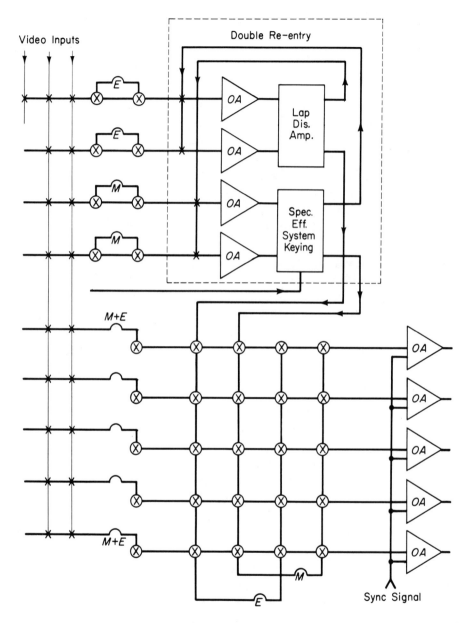

Time-Delay Equalization

COURTESY COHU ELECTRONICS, INC.

FIGURE 15-9. A "double re-entry" switching system.

routed through the mixer, a delay of *M* must be introduced to simulate the mixer delay time. It follows that any switchpoints that place video onto the output line without going through either the effects *or* the mixer must have a delay of *M* plus *E* to simulate both. Again, all of these conditions assure that the output will always maintain the same phase regardless of the switching paths taken. The automatic switch points serve the same purpose as previously stated.

Figure 15-9 illustrates the *double re-entry* switching system. This allows the mixer to feed the special effects generator *or* the special effects to feed into the mixer. This added versatility brings further complications, as the delay lines can no longer remain permanently affixed within the circuitry, but must be switched depending upon the situation with regard to the switcher operating configuration. Automatic crosspoints are now used to switch delay lines into and out of the video paths depending upon the delays that are required. Such complicated switching tasks are usually done with the help of electronic logic circuits.

## SWITCHING CIRCUITS

Figure 15-10 is a functional block diagram of a typical *building-block* video switching *card*. Physically it may take the form of a plug-in printed circuit card. It will accept one video input and provide five isolated video outputs. The video outputs may be switched so that only one output is being delivered, or all five. This particular type of switch card consists of an isolation amplifier, utilized primarily to achieve a high-input impedance, and five video switches, or crosspoints. The crosspoints are diode-isolated from the video output buss that is connected to each of them. These video outputs are each connected to an individual output amplifier card located nearby.

If it were desired to have 20 output possibilities for one video input, it would be necessary to have three more cards connected in parallel with the input. Therefore, the video would have to "loop-through" the first three cards and have the necessary 75-ohm termination affixed to the last card. With the high-input impedance of the first three cards and the proper termination of the last, it should be possible to turn on all 20 outputs with no noticeable effect on the input.

It may be observed from the diagram that the isolation amplifier has

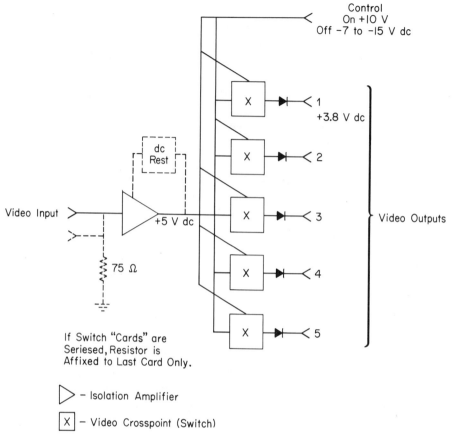

FIGURE 15-10. Block diagram of a video switch "card."

a dc restorer associated with it. It is the purpose of this element to sample the energy content of the video signal and derive a dc voltage which it then feeds back to the input to assure that blanking level is held constant. Consequently, when switching between two different video signals on separate switch cards, the blanking level will be consistent. If this were not done, it would be possible to generate large transients during the switch.

The dc output of the isolation amplifier is approximately 5 volts and the output of the video switch (crosspoint) is about 3.8 volts dc, with the video signal superimposed on each of these levels.

In the event it were desired to switch three inputs onto five possible outputs (3 × 5 system), it would merely be necessary to parallel the outputs of three switch cards. The five outputs of each card would then be in parallel with their counterparts on the other boards. However, each is effectively isolated from the other by the fact that when one of the parallel outputs is energized, the 3.8 volts that it generates back biases the diodes on the other boards. When diodes with extremely low capacitance are chosen for this purpose, the isolation is very effective.

An output amplifier card is illustrated in basic block form in Figure 15-11. It is a negative feedback type amplifier with the output taken from the collector. It is, therefore, a constant current source rather than a constant voltage source. This accounts for the 75-ohm resistor connected from the output to ground as a source termination. (A constant voltage source would have called for a resistor to be in series with the output rather than in parallel as shown.) The output amplifier also has an isolating diode connected to its output. As before, this is useful when it is desired to connect more than one output amplifier to a single video line.

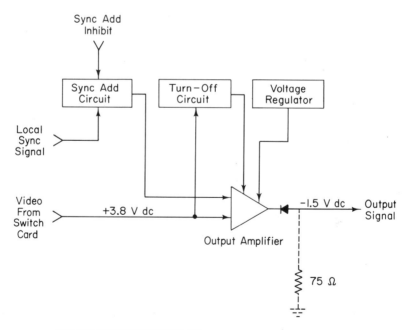

FIGURE 15-11. Block diagram of an output amplifier.

The diagram of Figure 15-11 depicts a turn-off circuit whose duty it is to detect the presence or absence of the 3.8 volts dc which is the dc output level of an energized video switch crosspoint. When there is not a switch point turned on at the input to this amplifier, the turn-off circuit will disable it.

In many cases, switching systems are called upon to switch composite and noncomposite video signals. Since the output of the switcher is normally expected to be composite, a means is provided in the output amplifier circuitry to add sync to the noncomposite video as needed. In the circuit shown, the local sync signal is always present at the input to the sync

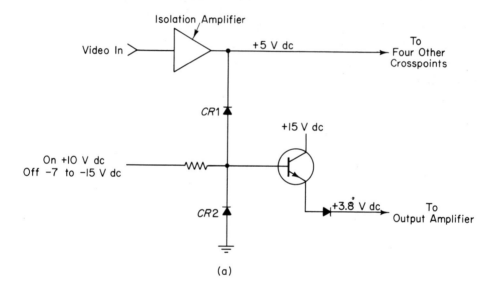

(a)

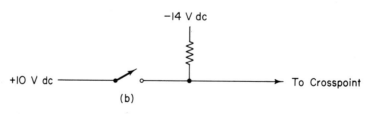

(b)

COURTESY COHU ELECTRONICS, INC.

FIGURE 15-12. A typical crosspoint and control circuit.

adding circuit; however, when this amplifier is called upon to pass a signal that already has the sync added to it, an inhibit signal from some form of external logic will prevent the sync adding circuit from functioning.

Figure 15-12 illustrates a simplified schematic representation of a typical crosspoint. The crosspoint is energized and allowed to pass video by applying a dc voltage to the junction of $CR1$ and $CR2$. A positive 10-volt potential at this point causes $CR2$ to become reverse biased so that it is electrically not in the circuit. Since the positive 10 volts is considerably higher in amplitude than the positive 5 volts on the video coming from the input isolation amplifier, $CR1$ is forward biased and this allows the passage of video into the emitter follower. In order to turn off the device, a negative voltage is routed to the junction of the two diodes. This will forward bias $CR2$, clamping the base of the transistor to ground (neglecting the few tenths of a volt drop across the diode). The action of $CR2$ reverse biases the base-emitter junction of the transistor and turns it off. With this crosspoint off, any other crosspoint connected to the same output amplifier will apply its own $+3.8$ volts to the cathode of $CR3$, thus reverse biasing it also and removing any internal capacity within this crosspoint from the video buss.

Figure 15-12B illustrates one method of supplying the positive and negative voltages necessary to activate and deactivate a crosspoint of the type described.

It has been previously mentioned that direct pushbutton control for activating the various crosspoints may not always be the best method. Figure 15-13A illustrates one other method that may be used, incorporating a flip-flop multivibrator. To understand its operation, assume that p-n-p transistors are being used with the collector resistors connected to the negative voltage. When, due to the operation of the multivibrator, the right-hand transistor is saturated, the line connected to the crosspoint will be at the positive potential and the crosspoint will be activated. With the right-hand transistor of the flip-flop turned off, its collector will be at the negative potential and the crosspoint will be turned off. In actual practice, instead of using a resistor as the load of the right-hand transistor, a relay may be used to provide a lamp indication of crosspoint activation by using one set of its contacts to couple a dc voltage to the bulb.

Figure 15-13B depicts a method whereby switching may be caused to occur during the vertical blanking interval only. By closing the switch, the multivibrator is in a configuration where it will change states upon the

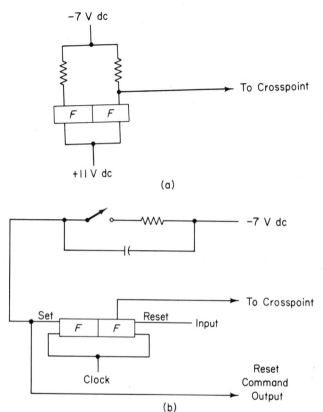

(a)

(b)

To Crosspoint

Set  Reset — Input

To Crosspoint

Clock

Reset
Command
Output

FIGURE 15-13. A multivibrator and vertical interval control.

application of a clock pulse. Generally, this clock pulse is derived from the vertical drive pulse output from a master sync generator. Thus, when the switch is depressed, the multivibrator is enabled and upon the application of the next vertical drive interval, the crosspoint will be activated, causing the switching to take place during the vertical blanking interval. The reset input provides an enabling voltage that will allow the flip-flop to return to its original state at the application of a clock pulse. This reset voltage is normally generated by the activation of a different crosspoint when it is desired to view a different scene. Thus, the action that activates a crosspoint also turns off the crosspoint that was previously activated.

In the event of a loss of vertical drive into the clock generator, it will free-run at approximately 30 hertz. Thus, switching may still be accomplished, although the advantages of the vertical interval switch will have been lost.

## DETAILED CIRCUIT DESCRIPTION

Figure 15-14 illustrates the complete schematic of a switch card of the type previously described. The isolation amplifier consists of two emitter followers, $Q1$ and $Q2$, connected in cascade, and $Q3$, which is part of a dc feedback path. The $Q2$ dc-output voltage is regulated with reference to the 5.5-volt supply and adjusted by potentiometer $R10$. The gain of this amplifier can be adjusted over a small range by potentiometer $R4$. The output of $Q2$ is applied to the five switching circuits.

Of the five identical switching circuits, only the $Q4$ circuit will be discussed. The voltages marked on the schematic are for $Q5$ switched off and $Q4$ switched on. The $Q4$ circuit, which includes $CR1$ and $CR6$, is controlled by signals that are fed to the switching matrix from external equipment.

A control signal is applied continuously to the switching circuit at either of two voltage levels: $+10$ volts when the switching circuit is to route the video signal to an output amplifier card, and $-14$ volts when the video signal is to be blocked. Since the cathode voltage of $CR1$ is $+5$ volts and the anode of $CR6$ is a zero potential, a control signal of $-14$ volts reverse biases $CR1$, removing the video, and forward biases $CR6$, which shunts the base of $Q4$ to ground. Also, the resulting voltage of $-0.3$ volt at the base switches $Q4$ off.

When the control signal is $+10$ volts, it reverse biases $CR6$ and forward biases $CR1$. The resulting voltage of approximately 5 volts at the base switches $Q4$ on, which routes the video signal through $R26$ and $CR16$ to the output amplifier card.

When $Q4$ is switched off, $CR16$ prevents the video output of a switching circuit on another card connected to the same output amplifier card from being present beyond $DR16$.

All or any one of the five switching circuits may be in the switched-on state at any one time. Whenever a switching circuit is switched off, networks as typically represented by $CR21$, $R41$, and $C14$ are switched on. These networks are compensating loads for the isolation amplifier.

All five switching circuits are identical except for the values of $R26$ through $R30$. These resistors equalize the high-frequency response of the switching circuits. The loss between the input of the isolation amplifier and the output of a switching circuit is approximately 2 to 3 decibels, depending on the video frequency.

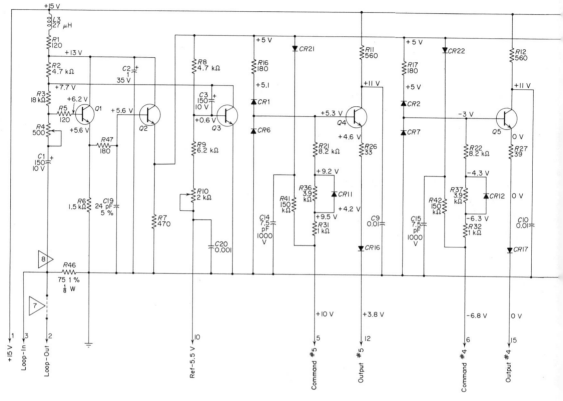

FIGURE 15-14. Circuitry of the switch "card" shown in Figure 15-10.

The dc restorer in the isolation amplifier, consisting of $Q9$, $R48$, $R50$, $C21$, and $C22$, provides the video signal with a constant blanking level regardless of the average picture level. The blanking level is adjusted by $R10$.

An example of an output amplifier is shown in Figure 15-15. $Q4$ and $Q5$ comprise a two-stage, direct-coupled amplifier with the output transistor, $Q5$, functioning as a current source. The video from the switch card input to the output amplifiers output is adjusted for an overall gain of unity by the low-frequency gain potentiometer, $R19$, and the high-frequency trimmer, $C3$.

The dc level of the video input from a switch card is usually positive, but it is approximately 0 volts when there is not a switched-on switching circuit on the line to the output amplifier. This latter condition should normally occur only in systems where more than one output amplifier feeds

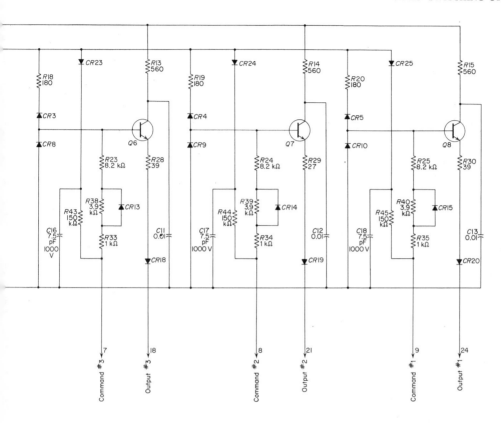

a single line, that is, where video output loop-through with amplifiers having turned-off circuits are used. When such an amplifier is turned off it is isolated from the common line by $CR10$.

The sync insertion circuit consists of two emitter followers, $Q1$ and $Q2$, connected in cascade, and a two-stage switch, $Q3$ and $CR14$. The sync signal of approximately 4 volts peak-to-peak is continuously fed to $Q1$, and when the switch is closed, the sync signal is applied through $CR14$ to $Q4$ of the output amplifier. The sync signal is adjusted to 0.3 volt peak-to-peak at the output of the amplifier by potentiometer $R8$ (it is adjusted to 0.4 volt if the system is being used for closed circuit purposes). The mixing of the sync signal with the video signal is inhibited when a positive voltage is applied to the base of $Q3$. The positive voltage causes $Q3$ to saturate, shunting the sync signal to ground, which reverse biases $CR14$ and thereby opens the two-stage switch. The inhibit signal, which is the positive control

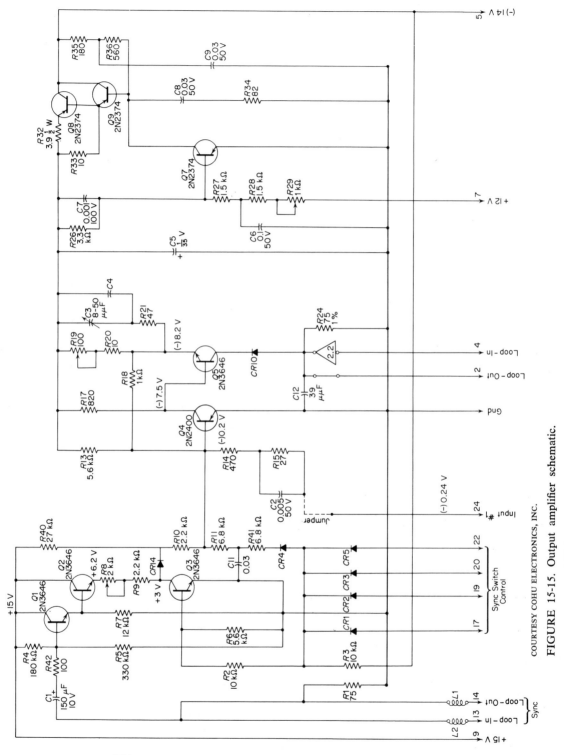

COURTESY COHU ELECTRONICS, INC.

FIGURE 15-15. Output amplifier schematic.

voltage used for switching on switching circuits having composite inputs, is applied to the sync insertion circuit through *CR*1, *CR*2, *CR*3, or *CR*5.

The voltage regulator on the output amplifier, consisting of *Q*7, *Q*8, and *Q*9, provides regulated voltage at low impedance for the output amplifier. The *Q*7 circuit is a regulated error amplifier, and *Q*8 and *Q*9 are emitter-follower passing transistors. The output voltage of the passing transistors is regulated with reference to the +12-volt supply. Potentiometer *R*29 is used to adjust the dc level of the output amplifier's output.

The turn-off circuit functions as follows: The dc level of the video input of the output amplifier is applied through *R*39 and *CR*13 to *Q*10, causing it to turn off. The effect of *Q*10 turning "off" is to permit the voltage regulator to function, thus energizing the amplifier. If all the switching circuits on the same buss to an amplifier are in the switched-off state, the dc input of the amplifier is zero, and *Q*10 switches on, causing the voltage regulator to turn off. This, in turn, de-energizes the amplifier.

# 16

## BASIC PRINCIPLES OF
## VIDEO TAPE RECORDING

Video tape recorders have become an increasingly important part of many television systems. Continued cost reduction and simplified operating and maintenance procedures will undoubtedly make them even more attractive. Continual advancement in the state of the art has resulted in a reduced size, complexity, and cost to the extent that home television tape recorders are a current reality. Because the video tape recorder is a complex device, worthy of several volumes in itself, no attempt will be made to give a detailed analysis, but an examination of the basic principles employed can be quite helpful in understanding present techniques.

Anyone who is familiar with audio tape recorders is aware that achievement of a specific frequency response depends largely upon the speed at which the tape passes the record and reproduce heads. Because the magnetic patterns that are established on the metal oxide that coats recording tape occupy a definite amount of space, only a certain number of such patterns may be formed in a given length of tape with any degree of fidelity. It is a well established fact that audio tape recorders that employ tape movement of 7½ or 15 inches per second (in/s) do a much better job of reproducing high audio frequencies than do tape recorders that might operate at 3¾ in/s.

Obviously, the frequencies encountered in a television video signal are of a much higher order than any audio recorder could hope to reproduce. Not only is the tape speed of an audio recorder too slow, but head design, amplifier bandwidth characteristics, and drive stability are completely inadequate. Many video tape recorders must possess the ability to record and

FIGURE 16-1. A broadcast video tape recorder.

accurately reproduce color television signals, including the 3.58-megahertz subcarrier component, and such units employ tape-to-head velocities of over 1000 inches per second.

The high tape velocities necessary for proper video reproduction on a standard transport mechanism give rise to several problems. The large bulk of the tape necessary for a recording of any appreciable duration is, in itself, a discouragement. Couple to this the difficulties that might be encountered with tape spill, speed regulation, and braking, and the direct method of recording becomes even less attractive, its relative simplicity and low cost notwithstanding. The most widely used method of achieving the necessarily high tape-to-head speeds has been to employ rotating heads that scan the tape at a high velocity while it moves past the scanning mechanism at a reasonably low speed.

## SCANNING METHODS

There are several means by which the tape may be scanned by moving heads in a video tape recorder. The first (Figure 16-2) is to pull the tape around a cylindrical head assembly at an angle to the horizontal. A slot is provided in the cylinder for the head to come into contact with the tape as it rotates within the cylinder, providing an angular inscription of magnetic detail across the oxide-coated surface of the tape. When the next cycle of rotation begins, the tape has moved slightly forward (governed by the capstan speed) and the next inscription is impressed in a path that is parallel to the original, yet each is separate and distinct. Such action results

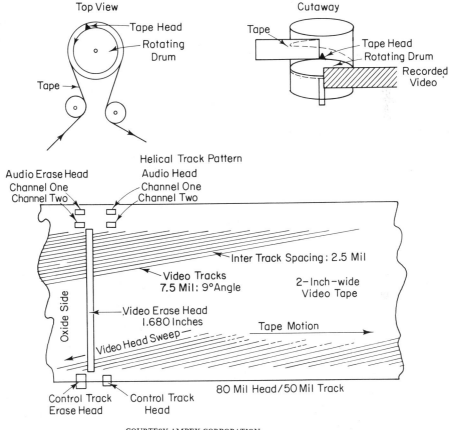

COURTESY AMPEX CORPORATION.

FIGURE 16-2. A single head helix scanning system.

in an increase in the storage capacity of the magnetic tape and yields a tape speed that is reasonable. Increasing the width of the tape, from the normal ¼ inch common to most audio recorders, to ½ to 2 inches also enhances the ability to record signals of several megahertz bandwidth.

Scanning a video tape in the above manner is referred to as a *helical* scanning method. Because the tape is formed in a nearly complete spiral around the rotating single head, it is sometimes referred to as a *full-helical* or *single-head helix* scanning system. It may be noted from the diagram that the head must leave the tape momentarily once during each cycle as it travels from the trailing edge of the tape to the leading edge where it will begin the succeeding scan. This, of course, causes a loss, or dropout, of signal and, without adequate measures being taken in signal processing, may cause an objectionable disturbance in the reproduced television picture. The rotational velocity of the record/reproduce head is normally governed to ascertain that this drop-out will occur during the vertical blanking interval of the video signal and thus be allowed to occur unobserved. Figure 16-3 illustrates a means of helical recording that uses two rotating heads, spaced 180 degrees apart, each sweeping the same arc in a repetitive manner. In this case, the tape need not be wrapped in a complete loop around the head assembly, and it should be noted that one head will be in contact with the tape at all times. Thus, if the heads are electrically switched, so that the head that is in contact with the tape is continuously active during its pass, the rather long signal drop-out that was present in the previous example is largely eliminated and the only disturbing occurrence that might be expected would be the rapid switching of the heads. Head switching may also be made to occur during vertical blanking when it will not be visible. This type of scanning is referred to as the *half-helical* or *two-head helix* method of video recording.

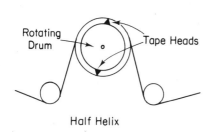

Half Helix

FIGURE 16-3. A two-head helix scanning system.

The method of recording used by a great many broadcast video tape recorders is known as the *transverse* method of scanning. Figure 16-4 illustrates the unidirectional path of the tape past a rotating headwheel which has 4 record/reproduce heads affixed to it at accurately spaced 90-degree intervals. The tape is formed into a concave surface by special tape guides that assure that each head will maintain a proper, fixed-pressure contact with the oxide surface. In this example, two other channels of information are also shown, both of which are also utilized when generating recording patterns with the full- and half-helical systems previously described. These additional channels are included to provide for recording of audio information and control signals. The audio capability is self-explanatory, but the control signal system is used to establish precise references along the length of the tape, to control the tape velocity past the scanning heads during the playback mode of operation.

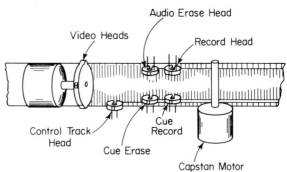

Basic Principle of Transverse Recording. Four—Head Video Drum, Longitudinal Audio Cue and Control Tracks. Tape Speed, 15 and $7\frac{1}{2}$ Ips; Audio Track Width, 10 and 15 Mils; Guardbands, 5 and $2\frac{1}{2}$ Mils. Control Track Also Contains Frame Pulses for Editing Purposes.

FIGURE 16-4. The transverse recording system.

In the transverse system depicted in Figure 16-5, the rotating head will revolve at a speed that will allow it to make four complete rotations during the period of one vertical field. Thus, sixteen complete video inscriptions will be developed and situated in a parallel manner in a near perpendicular direction across the tape. In a typical system the vertical sync will be recorded by video head number one at a precisely determined spot near the center of the tape width. In the *record* mode of operation, the servo system tightly locks the motor velocity and phase to the vertical sync of the incoming video waveform, and a rigid phase relationship is established, enabling the recording of the vertical to be accurately maintained as to position on the tape. If the headwheel velocity were fixed, and no provisions were made for slaving it to the vertical sync, the recorded vertical sync pulses would have no fixed position on the tape and would drift across the recorded tracks. Although a tape recorded under such conditions might be successfully played back on the same machine (although no splices could be made after the recording were accomplished), the tape

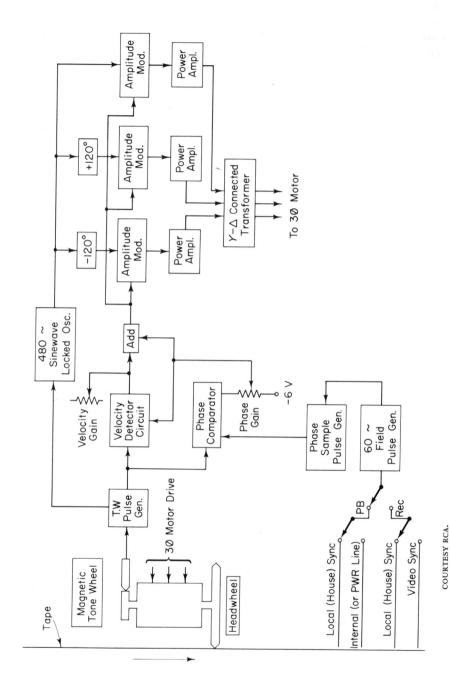

COURTESY RCA.

FIGURE 16-5. Servo system in RCA TR-70 video tape recorder.

would certainly not be interchangeable with tapes made on a different machine. Thus, a servo system is needed that will control the motor speed as a function of the incoming video signal synchronizing pulses.

In the system shown, a magnetic tone wheel is located on the head-wheel motor, providing an output waveform that accurately represents the speed at which the headwheel is rotating. When recording, this output may be compared with the incoming video sync signal, and an error voltage can be developed that will govern the motor speed to assure correct place-ment of the signals on the tape inscription. During playback, the primary function of the headwheel and capstan servo systems is to assure a stable picture by properly regulating the motor speeds so that tape movement past the heads and the angular velocity of the head assembly remain strictly controlled and accurate. The capstan servo system samples the control channel recorded along the edge of the tape and keeps the tape moving at a speed that will assure precise alignment of each recorded path with the scanning heads.

## THE VIDEO SYSTEM

Figure 16-6 illustrates a simplified block diagram of the video system in a video tape recorder. Probably the first thing that will be noted is that the video signal is not directly recorded onto the tape, but is first converted into a frequency-modulated signal and then recorded.

FM recording is advantageous in that it limits the spectrum of fre-quencies that must be handled by the tape heads and their efficiency, there-fore, remains relatively constant. Also, because the amplitude of the FM signal remains constant throughout its frequency deviations, any amplitude irregularities that might occur during playback due to varying tape-to-head distance, tape irregularities, etc., may be compensated for by AGC circuits. Therefore, when the signal is demodulated into an amplitude-modulated waveform, the irregularities that might be caused by the aforementioned problems will not be disturbingly evident.

Professional tape recorders, such as those used in broadcast for re-cording color signals, will incorporate special circuitry to pre-emphasize the video signal that is to be recorded in such a manner as to overcome any undesirable effects that may result from the recording process. When the signal is reproduced during playback, processing circuitry will typically regenerate tape sync, add new color burst, establish a new and constant black level, perform gamma correction and aperture compensation, as well as amplify the video signal to the correct output level. In addition to all of this, the video signal must be processed so the various sync, blanking, and subcarrier components may be used to govern the tape speed, head velocity, etc.

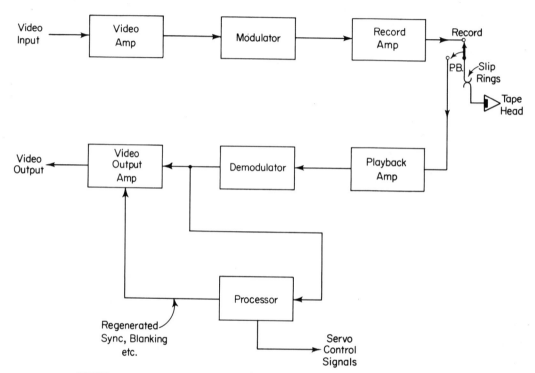

FIGURE 16-6. Video signal is FM modulated to narrow the bandwidth required of the record/reproduce heads.

Another desirable feature in a broadcast tape machine that serves to further complicate matters is an ability to cause the playback of a recorded program to occur with its sync and subcarrier waveforms locked to the phase of the station or network master-sync generator system. This will allow video switching to be accomplished from the tape recorder signal to other synchronous video information without the disturbances caused when switching between two nonsynchronous video signals.

A block diagram of a professional television tape recorder (RCA TR-70) is shown in Figure 16-7. Although machines used in closed circuit applications are typically not so complex, this illustration serves to show

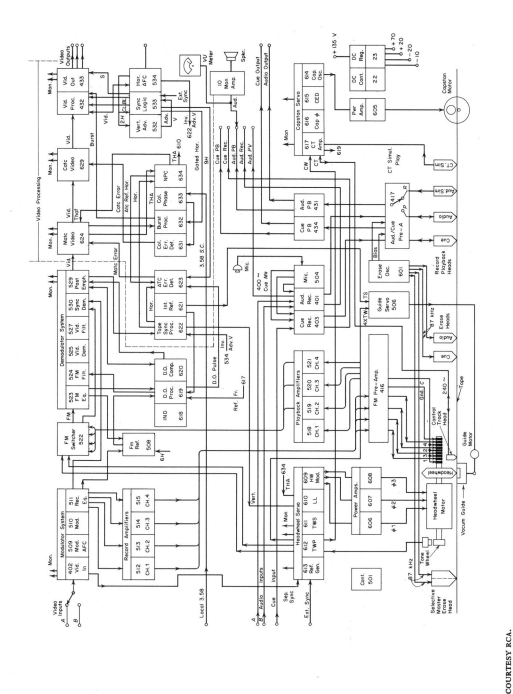

FIGURE 16-7. Functional block diagram of the RCA TR-70 professional tape recorder.

the rather awesome amounts of circuitry and signal processing that are involved in video tape recording.

While this short discussion of television tape recorders cannot hope to more than outline the broad concepts of such systems, it has been included to indicate the fascinating diversification of equipment that will be encountered by the individual who becomes actively interested or involved in the field of television.

# 17

# GLOSSARY OF
# TELEVISION TERMS

The field of television, in common with many specialized disciplines, has many technical, scientific, and slang terms that are used to describe equipment, methods, concepts, and phenomena associated with it. To a person not familiar with these expressions, a conversation concerning the subject can be an exercise in confusion. The following terms and expressions encompass a great many of those in common usage today.

The author is indebted to the Electronic Industries Association for many of the definitions used. These are so indicated (EIA).

ABERRATION. An optical defect that prevents all of the light rays processed by the lens from being brought to the same focal plane. A defect in a cathode-ray tube, in which the electron lens (focus field) does not bring the electron beam to the same point of sharp focus at all points on the screen:

SPHERICAL. Due to the spherical form of a lens or mirror, central and marginal rays from a point source on the axis converge to a different point of focus.

CHROMATIC. Due to variation of refractive index with wavelength in a refractive medium, each wavelength of light energy has its own distinct point of focus.

ASTIGMATISM. Rays from a point source are not brought to a single focal point, due to improper convergence within the lens system. Generally refers to a point source on the axis of the lens.

COMA. Similar to astigmatism, but pertains to a point source not on the axis of the lens.

AMBIENT TEMPERATURE. The temperature of the medium (air, water, etc.) that surrounds and contacts an object.

AMBIENT LIGHT LEVEL. The intensity of light surrounding an object.

APERTURE. An opening that permits light, electrons, or other forms of radiation to pass through. In an electron gun, the aperture determines the size, and has an effect upon the shape, of the electron beam. In television optics, it is the effective diameter of the lens that controls the amount of light reaching the camera tube.

APERTURE CORRECTION. In television, a means whereby a video signal is electronically enhanced to increase image sharpness.

APERTURE MASK. A metal plate with accurately formed holes, placed close behind the phosphor-dot faceplate in a tricolor picture tube. It insures that each of the three electron beams excites only the desired phosphor dot. Also called a *shadow mask*.

ASPECT RATIO. The ratio of width to height for the frame of the televised picture. The U.S. standard is 4:3.

AUTOMATIC FREQUENCY CONTROL (AFC). A method whereby the frequency of an oscillator is automatically controlled by additional circuitry.

AUTOMATIC GAIN CONTROL. A process by which gain is automatically adjusted as a function of input or other parameter. (EIA).

AUTOMATIC TARGET CONTROL. Automatic adjustment of vidicon target potential as a function of scene brightness.

BACK PORCH. That portion of a composite picture signal that lies between the trailing edge of a horizontal-sync pulse and the trailing edge of the corresponding blanking pulse. *Note:* The color burst, if present, is not considered a part of the back porch. (EIA).

BANDWIDTH. The range of frequencies within which performance, with respect to some characteristic, falls within specific limits. (EIA).

BARREL DISTORTION. Distortion that makes the televised image appear to bulge outward on all sides like a barrel.

BEAM. A concentrated, unidirectional flow of electrons or other energy.

BEAM POSITIONING MAGNETS. Three small magnets placed at 120-degree intervals around the neck of a tricolor picture tube. Permits control of the angle at which each beam approaches the aperture mask. Also, magnets used in proximity of a vidicon tube for beam centering.

BEAM WIDTH. Angular width of a beam, measured in azimuth.

BLACK COMPRESSION. The reduction in gain applied to a picture signal at those levels corresponding to dark areas in a picture, with respect to the gain at that level corresponding to the midrange light value in the picture. (EIA). Reduces contrast in the low lights of the picture as seen on a monitor.

BLACK LEVEL. That level of the picture signal corresponding to the maximum limit of black peaks. (EIA).

BLACK NEGATIVE. Television picture signal in which the polarity of the voltage corresponding to black is negative with respect to that which corresponds to the white area of the picture tube.

BLACKER-THAN-BLACK. The amplitude region of the composite video signal below reference black level in the direction of the synchronizing pulses.

BLANKING. The process of cutting off the electron beam in a camera tube or picture tube during retrace.

BLANKING LEVEL. That level of a composite picture signal which separates the range containing picture information from the range containing synchronizing information. *Note:* The "setup" region is regarded as picture information. (EIA).

BLANKING SIGNAL. A wave constituted of recurrent pulses, related in time to the scanning process, used to effect blanking. (EIA).

BLEEDING WHITE. An overloading condition in which white areas appear to flow irregularly into black areas. (EIA).

BLOOMING. The defocusing of regions of the picture tube brightness at an excessive level, due to enlargement of beam spot size.

BOUNCE. Sudden variations in picture presentation (brightness, size, etc.) independent of scene illumination. (EIA).

BBEATHING. Variations similar to bounce but at a slow, regular rate. (EIA).

BRIGHTNESS. The attribute of visual perception in accordance with which an area appears to emit more or less light. (EIA).

BURN. In a vidicon or other camera tube, a burn is the sticking or persistence of an image in the output signal of a television camera which is visible on the monitor screen after the camera has been turned away from the image. In a picture tube it is a blemish or localized discoloration of the screen.

CAMERA CHAIN. A television camera, associated control units, monitor, associated equipment, and connecting cables.

CAMERA CONTROL UNIT. A unit, generally removed from the camera head, which allows remote operation of camera functions (such as focus, beam, width, height, centering, etc.). In closed circuit television it may commonly contain power supplies, sync generator, video processing circuits, etc.

CAMERA HEAD. A camera tube, deflection system, and a minimum of components, generally located some distance from a camera control unit and connected by multiconductor cabling.

CAMERA TUBE. An electron tube for the conversion of an optical image into an electrical signal by a scanning process. (EIA). Also called a pickup tube.

CANDLEPOWER. Refers to the amount of light in a given area. One foot-candle is the amount of light emitted by a standard candle at one foot distance. Average office or classroom lighting generally ranges from 25 to 75 foot-candles. Bright sunlight is approximately 10,000 foot-candles. The desirable light level in a television studio is generally a minimum of about 250 foot-candles for monochrome and 400 foot-candles for color.

CATHODE-RAY TUBE (CRT). A tube in which the electrons emitted by a heated cathode are focused into a beam and directed toward a phosphor-coated surface that may then become luminescent at the point where the electron beam strikes it.

CCTV. Closed circuit television.

CHROMATIC ABERRATION. See "Aberration."

CHROMATICITY. The color quality of light that is defined by the wavelength (hue) and saturation. Chromaticity defines all the qualities of color except its brightness.

CHROMINANCE. A color term defining the hue and saturation of a color. Does *not* refer to brightness.

CHROMINANCE SIGNAL. The *I* and *Q* sidebands of the color subcarrier in a modulated color signal. These carry information relative to hue and saturation of colors but none relative to brightness.

CIRCLE OF CONFUSION. The image of a point source that appears as a circle of finite diameter because of the aberrations inherent in an optical system (including the eye).

CLAMPING. The process that establishes the picture black level at a predetermined reference potential at the beginning of each scanning line.

CLEAR APERTURE. Area of an optical element having no defects or obstructions to interfere with the transmitted or reflected radiation. Usually circular and expressed in terms of its diameter.

CLIPPING. The removal of the peaks of a signal.

CLOSED CIRCUIT TELEVISION. A television system wherein all video information is carried by wire circuitry or microwave relays for specialized audiences. A tool widely used in education, industry, and science.

COAXIAL CABLE (COAX). A conductor centered inside of and insulated from a cylinder that is also used as a conductor. The two conductors are separated by a dielectric. Commonly used to carry video signals.

COLLECTOR (OPTICAL). Any lens or mirror that collects or converges radiation.

COLOR BURST. In NTSC color, normally refers to a burst of approximately 8 or 9 cycles of 3.58-megahertz subcarrier on the back porch of the composite video signal. Serves as a color synchronizing signal to establish a frequency and phase reference for the chrominance signal.

COLOR DECODER. In color television reception, an apparatus for deriving the receiver primary signals from the color picture signal and the color burst. (EIA).

COLOR ENCODER. A device that produces a NTSC color signal (encoded) from separate *R, G,* and *B* video inputs. May also generate the color burst. Also known as *color modulator,* and *colorplexer.*

COLOR PURITY. A term used in reference to the operation of a tricolor picture tube. It refers to the production of pure red, green, or blue illumination of the phosphor-dot faceplate.

COLOR SUBCARRIER. In NTSC color, the carrier whose modulation sidebands are added to the monochrome signal to convey color information; i.e., 3.579545 megahertz.

COLOR SUBCARRIER OSCILLATOR. A stable 3.58-megahertz (actually 3.579545 megahertz) oscillator used to generate the color subcarrier.

COMA. A defect in a cathode-ray tube that makes the normally circular electrom beam appear comet shaped at the edges of the tube screen. See "aberration."

COMPATIBLE COLOR. The characteristic of a color video signal that makes it acceptable for use in a monochrome receiver.

COMPLEMENTARY COLOR. A color formed by subtracting a sample color from white light. For example, if red is subtracted, leaving blue and green, the complementary color is cyan (blue-green).

COMPOSITE PICTURE SIGNAL. The signal that results from combining a blanked picture signal with the sync signal. (EIA).

COMPRESSION. The reduction in gain at one level of a picture signal with respect to the gain at another level of the same signal. (EIA).

CONCAVE. Curved inward.

CONTRAST. The range of light and dark values in a picture, or the ratio between the maximum and minimum brightness values. (EIA). Sometimes expressed as contrast ratio.

CONVERGENCE. The crossover of the three electron beams in a color television picture tube. This normally occurs at the plane of the aperture mask.

CONVEX. Curved outward.

CRT. Cathode-ray tube.

CYAN. The complement of red (blue-green).

DARK CURRENT. The current that flows in a photoconductor when it has no illumination impressed upon it.

DECODER. The circuitry in a color television receiver that transforms the detected color signals into a form suitable to operate the color tube being used.

DEFINITION. The fidelity with which a televised image represents the original scene.

DEFLECTION. A term used to denote beam movement across a pickup tube or display tube.

DEPTH OF FIELD. The "in focus" range of an optical system, measured from the distance behind an object to the distance ahead of an object where the apparent focus is maintained.

DETAIL. The smallest elements resolvable on a monitor display. Similar to resolution.

DETAIL CONTRAST. The ratio of the amplitude of video signal representing high-frequency components with the amplitude representing the reference low-frequency component, usually expressed as a percentage at a particular line number. (EIA).

DICHROIC MIRROR. A type of front-surfaced color filter that passes certain colors, but reflects others.

DIFFERENTIAL GAIN. The amplitude change introduced by the overall circuit, measured in decibels or percentage, as the picture signal undergoes excursions from blanking (black level) to white level.

DIFFERENTIAL PHASE. The phase change of a signal (such as the 3.58-megahertz color subcarrier) introduced by the overall circuit, measured in degrees as the picture signal undergoes excursions from blanking (black level) to white level.

DIFFRACTION. The apparent deflection of rays of light as they pass by sharp edges, producing multicontrast, or multihued, bands of light when projected onto a surface.

DISPERSION. The variation in angular travel for various colors of light in a medium due to the different index of refraction for the various wavelengths.

DISPLACEMENT OF PORCHES. Refers to any difference between the level of the front porch and the level of the back porch. (EIA).

DISTORTION. Any deviation of a waveform or optical image from its original shape (or phase).

DOLLY. A frame or other structure equipped with wheels on which the tripod or pedestal supporting television cameras or other equipment is mounted.

DRIVE. Generally, a signal that must be applied to circuitry for it to operate properly (such as horizontal drive, vertical drive, etc.).

DYNAMIC CONVERGENCE. A means whereby the three beams in a color television picture tube are caused to maintain correct convergence over the entire face of the tube by use of electromagnetic fields modulated by waveforms occurring at the horizontal and vertical rates.

ECHO. A reflection of a transmitted television picture, appearing as a "ghost" on the screen of a monitor or receiver.

EIA. Electronic Industries Association.

EIA SYNC. The signal employed for the synchronizing of scanning specified in EIA Standards publications.

ELECTROMAGNETIC FOCUSING. A method of focusing a cathode-ray beam to a fine spot by application of electric currents through a coil placed around the tube.

ELECTROSTATIC FOCUSING. A method of focusing a cathode-ray beam to a fine spot by use of electrostatic potentials.

ENCODER. A device in color television which produces an NTSC color signal from separate red, green and blue video inputs.

EQUALIZER. Compensates for undesirable losses or distortions of certain frequencies within a television system.

EQUALIZING PULSES. Pulses that occur in the standard television signal at twice the horizontal line frequency, occurring just before and just after the vertical synchronizing pulses. They minimize the effect of horizontal sync pulses on the interlace.

ETV. Educational television.

FADE. To change signal strength.

FADE IN. To increase signal strength gradually. Opposite of fade out.

FADE OUT. To decrease signal strength gradually. Opposite of fade in.

FADER. A control or group of controls for effecting fade in and fade out of video or audio signals.

FIELD. One of the equal parts into which a television frame is divided in an interlaced system of scanning. One vertical scan, containing many horizontal scanning lines, is generally termed a field.

FIELD FREQUENCY. The product of frame frequency multiplied by the number of fields contained in one frame. (EIA). The U.S. standard is 60 fields per second. Also called field repetition rate.

FIELD LENS. Lens used to effect the transfer of the image formed by an optical system.

FIELD OF VIEW. The solid angle that an optical system can see.

FILM CHAIN. An equipment arrangement in which one or more film projectors are directed in turn to provide image pickup for a television camera.

F/NUMBER. Relative aperture. Ratio of diameter to focal length of a lens or mirror.

FOCAL LENGTH. Distance from the principal point in a lens to the actual focal point.

FOCAL POINT. The point at which a lens or mirror will focus parallel light rays.

FOCUS. The point at which light rays or an electron beam form a minimum-size point or spot. Also, the action of bringing light or electrons to a fine spot.

FOCUS CONTROL. A manual adjustment for bringing the electron beam of a camera pickup tube or picture tube to a minimum-size spot, producing the sharpest image.

FOOT-CANDLE. A unit of illuminance when the foot is taken as the unit of length. It is the illuminance on a surface one square foot in area on which there is a uniformly distributed flux of one lumen, or the illumination at a surface all points of which are at a distance of one foot from a uniform source of one candle. (EIA).

FOOT-LAMBERT. A unit of luminance equal to $1/\pi$ candle per square foot, or to the uniform luminance of a perfectly diffusing surface emitting or reflecting light at the rate of one lumen per square foot. (EIA).

FRAME. The total area, occupied by the picture, which is scanned while the picture signal is not blanked. (EIA). Television in the United States commonly employs two interlaced fields per frame.

FRAME FREQUENCY. The number of times per second that the complete frame is scanned. The U.S. standard is 30 frames per second.

FREE-RUNNING FREQUENCY. The frequency at which a normally synchronized oscillator operates in the absence of a synchronizing signal. (EIA).

FREQUENCY INTERLACE. In color television, the method by which color and black and white sideband signals are interwoven within the same channel bandwidth.

FRONT PORCH. That portion of a composite signal that lies between the leading edge of the horizontal blanking pulse and the leading edge of the corresponding sync pulse. (EIA).

F/STOP. Refers to the speed or relative ability of a lens to pass light. It is calculated by dividing the focal length of the lens by its diameter.

GAMMA. In television, the exponent of that power law which is used to approximate the curve of output magnitude vs. input magnitude over the region of interest. (EIA).

GAMMA CORRECTION. The introduction of a nonlinear output-input characteristic for the purpose of changing the effective value of gamma. (EIA).

GEOMETRIC DISTORTION. Any aberration that causes the reproduced picture to be geometrically dissimilar to the perspective plane projection of the original scene. (EIA).

GLITCH. A form of interference appearing as a horizontal bar or picture offset moving vertically through the picture. It may be stationary when the vertical scan is phase locked to the power line frequency.

GRAY SCALE. Variations of value from white, through shades of gray, to black on a television screen. The gradations approximate the tonal values of the original image picked up by the television camera.

HALATION. A glow or diffusion that surrounds a bright spot on a television picture tube screen. A defect in quality is indicated.

HIGH-CONTRAST IMAGE. A picture in which strong contrast between light and dark areas is visible. Intermediate values, however, may be missing.

HIGH DEFINITION. High resolution. The television equivalent of high fidelity.

HIGHLIGHTS. The maximum brightness of the picture, which occurs in regions of highest illumination. (EIA).

HORIZONTAL BLANKING. The blanking signal at the end of each scanning line. (EIA). Allows horizontal retrace to occur unobserved.

HORIZONTAL (HUM) BARS. Relatively broad horizontal bars, alternately black and white, which extend over the entire picture. They may be stationary or may move up or down. Sometimes referred to as a "venetian-blind" effect, it is caused by an approximate 60-hertz interfering frequency or one of its harmonic frequencies. (EIA).

HORIZONTAL INTERVAL. The period of time required to complete one horizontal scan and retrace.

HORIZONTAL RESOLUTION. The number of individual picture elements that can be distinguished in a horizontal scanning line within a distance equal to the picture height.

HORIZONTAL RETRACE. The return of the electron beam from the right to the left side of the raster after the scanning of one line. (EIA).

HUE. The dominant color of an object as determined by the wavelength of the emitted or reflected light. It is the redness, blueness, greenness, etc., of an object.

HUM. Electrical disturbance at the power supply frequency or harmonics thereof. (EIA).

HUM MODULATION. Modulation of a radio frequency or detected signal by hum. (EIA).

ICI. International Commission of Illumination, or CIE, Commission Internationale d'Eclairage.

IMAGE DISSECTOR TUBE. A camera tube in which an electron image produced by a photoemitting surface is focused in the plane of a defining aperture and is scanned past that aperture. (EIA).

IMAGE ORTHICON. A camera tube in which an electron image is produced by a photoemitting surface and focused on one side of a separate storage target which is scanned on its opposite side by an electron beam, usually of low-velocity electrons. (EIA).

INCIDENT LIGHT. The light that falls directly on an object.

INDEX OF REFRACTION. The ratio of the velocity of light in a vacuum to the velocity of light in a refractive material for a particular wavelength of light.

INTERLACED SCANNING. A scanning process in which the distance from center to center of successively scanned lines is two or more times the nominal line width, and in which the adjacent lines belong to different fields. (EIA).

ION. A charged atom. In a cathode-ray tube it may be an atom of residual gas. (EIA).

ION SPOT. A spot on the fluorescent surface of a cathode-ray tube, which is somewhat darker than the surrounding area because of bombardment by negative ions that reduce the sensitivity. (EIA).

ION TRAP. An arrangement of magnetic fields and apertures that will allow an electron beam to pass through but will obstruct the passage of ions. (EIA).

IRE. Institute of Radio Engineers.

IRIS. An adjustable aperture built into each camera lens to permit control of the amount of light passing through the lens.

I SIGNAL. The color sidebands produced by modulating the color sub-carrier at a phase 57 degrees removed from the "burst," reference phase (sometimes known as the *in-phase* signal). This signal is capable of reproducing the range of colors from orange to cyan.

ISOLATION AMPLIFIER. An amplifier with input and output circuitry so designed to eliminate the effects of changes made at either upon the other.

JEEP. Modification of a television receiver intended for RF systems to accommodate a composite video signal, and possibly audio, by direct connection.

JITTER. Small rapid variations in a waveform due to mechanical disturbances or to changes in the supply voltages, in the characteristics of components, etc. (EIA).

KINESCOPE. Television picture tube.

KINESCOPE RECORDER. A film recording made by a 16-millimeter motion picture camera especially designed to photograph a television picture tube. The sound portion of the program is also recorded simultaneously. The film so produced can then be shown on any 16-millimeter projector, as well as on television.

LAG. In camera tubes, a persistence of the electrical charge image for a small number of frames. (EIA).

LENS SHAPES:

PLANO-CONVEX.   One convex side, one flat side.

DOUBLE CONVEX (BICONVEX).   Both sides convex.

PLANO-CONCAVE.   One concave side, one flat side.

DOUBLE CONCAVE (BICONCAVE).   Both sides concave.

MENISCUS.   One convex side, one concave side.

LENS SPEED. Refers to the ability of a lens to pass light. A fast lens might be rated f/1.4; a much slower lens might be designated as f/8. The larger the f/number, the slower the lens.

LIGHTING:

EYE LIGHT. A special source of illumination designed to effect desirable reflection from the eyes of a subject without substantially affecting the overall lighting condition.

FRESNEL. A special lens with concentric circle forms impressed in its front surface to focus spotlight beams for use in studio lighting. May be obtained in a variety of designs with restricted focusing from a 16-degree beam to a flood beam of some 70 degrees.

FILL LIGHT. Auxiliary illumination to lessen contrast range or to reduce shadows.

KEY LIGHT. The lighting effect indicating the direction of the major source of illumination of a scene.

SCOOP. A floodlight employed to illuminate large areas at close range.

SET LIGHT. Auxiliary illumination of the background or set, in addition to the lighting supplied for the major subjects or areas.

SIDE BACK-LIGHT. Off-center illumination behind the subject.

LINE AMPLIFIER. An amplifier for audio or video signals that feeds a transmission line; also called *program amplifier*.

LINE FREQUENCY. The number of horizontal scanning lines per second. Also sometimes used to refer to power line frequency.

LOAD. That which receives the output of a unit of equipment.

LOOP-THRU. The method of feeding a series of high impedance circuits (such as video monitors) from a pulse or video source with a coaxial transmission line, in such a manner that the line is terminated in its characteristic impedance at the last circuit input on the line.

LOSS. Power dissipation that serves no useful purpose.

LUMEN. Unit of light flux. It is the power of light falling on 1 square meter of a hollow sphere of 1-meter radius at the center of which is a light source of 1 candlepower.

MATRIX (SWITCHER). A combination or array of electromechanical or electronic switches that route a number of signal sources to one or more destinations.

MEGAHERTZ (MHz). One million cycles per second.

MESHBEAT. A characteristic of camera pickup tubes that may be visible on a monitor presentation as a series of closely spaced lines or a screen-like pattern. It generally is not visible unless the beam is de-focused slightly.

MOIRE. A wavy or satiny effect produced by convergence of lines. Usually appears as a curving of the lines in the horizontal wedges of the test pattern and is most pronounced near the center where the lines form-ing the wedges converge. A moire pattern is a natural optical effect when converging lines in the picture are nearly parallel to the scanning lines. To a degree this effect is sometimes due to the characteristic of color picture tubes and of pickup tubes (in the latter termed *mesh-beat*). (EIA).

MONITOR (VIDEO). A television viewer connected directly to the camera system output. A true video monitor does not include channel selector components.

MONOCHROME. Pertaining to black and white television systems, one chro-maticity.

MONOCHROME SIGNAL. (1) In monochrome television, a signal wave for controlling the luminance values in that picture. (2) In color televi-sion, that part of the signal wave which has major control of the lumi-nance values of the picture, whether displayed in color or in mono-chrome. (EIA).

MONOSTABLE. Having only one stable state.

MULTIPLEXER (OPTICAL). A specialized optical device that makes it pos-sible to use a single television camera in conjunction with one or more motion picture projectors and/or slide projectors in a film chain.

NAB. National Association of Broadcasters.

NAEB. National Association of Educational Broadcasters.

NARTB. National Association of Radio and Television Broadcasters. (Now the NAB.)

NEGATIVE IMAGE. Refers to a picture signal having a polarity that is oppo-site to normal polarity and which results in a picture in which the white areas appear as black and vice-versa. (EIA).

NTSC. Abbreviation for National Television System Committee, which established the standards for black and white and color television.

OBJECTIVE (LENS). The element or elements of an optical system that form(s) an image of the object.

OPTICAL AXIS. A straight line passing through the centers of the curved surfaces of a lens.

OVERSHOOT. An effect that may occur within a system with a change in signal amplitude, causing a momentary signal excursion beyond the steady-state level that would have been achieved if no distortion was present.

PAIRING. A partial or complete failure of interlace in which the scanning lines of alternate fields do not fall exactly between one another, but tend to fall (in pairs) one on top of the other. (EIA). Severely reduced vertical resolution capability results.

PEAK-TO-PEAK VOLTAGE. The sum of the extreme negative and positive alternations of a signal.

PERSISTENCE. The period of time a phosphor continues to glow after the excitation source is removed.

PHOTOCATHODE. An electrode used for obtaining photoelectric emission. (EIA).

PHOTOCONDUCTOR. A device in which electrical resistance varies in relationship with exposure to light.

PICTURE ELEMENT. Any segment of a scanning line, the dimension of which (along the line) is exactly equal to the nominal line width.

PICTURE TUBE (KINESCOPE). A cathode-ray tube used to produce an image by variation of the beam intensity as the beam scans a raster. (EIA).

PIN-CUSHION DISTORTION. Distortion in a television picture that makes all sides appear to bulge inward.

POLARITY OF PICTURE SIGNAL. The sense of the potential of a portion of the signal representing a dark area of a scene relative to the potential of a portion of the signal representing a light area. Polarity is stated as *black negative* or *black positive*. (EIA).

PREAMPLIFIER. An amplifier, the primary function of which is to raise the output of a low-level source to an intermediate level so that the signal may be further processed without appreciable degradation on the signal-to-noise ratio of the system. (EIA).

PRIMARY COLORS. Three colors wherein no mixture of any two can produce the third. In color television these are the additive primary colors, red, blue, and green.

PROJECTION TELEVISION. A unit composed of a high-intensity picture tube and a series of mirrors and lenses arranged in such a manner as to project an enlarged television image.

PULSE RISE TIME. The interval between the instants at which the instantaneous amplitude first reaches specified lower and upper limits, namely 10 per cent and 90 per cent of the peak-pulse amplitude unless otherwise stated. (EIA).

Purity Coil. An electromagnetic device placed about the neck of a three-gun tricolor picture tube. Its function is to control the angle at which all three beams approach the aperture mask. Its correct adjustment produces pure colors of red, green, and blue on the phosphor-dot faceplate.

Q Signal. The color sidebands produced by modulating the color subcarrier at a phase 147 degrees removed from the "burst" reference phase (sometimes known as the *quadrature* signal). This signal is capable of reproducing the range of colors from purple to yellow-green.

Random Interlace. In random interlace there is no fixed relationship between adjacent lines and successive fields. (EIA). The spacing between the lines varies randomly.

Raster. A predetermined pattern of scanning lines which provides substantially uniform coverage of an area. (EIA).

Raster Burn (Camera Tubes). A change in the characteristics of that area of the target which has been scanned, resulting in a spurious signal corresponding to that area when a larger or tilted raster is scanned. (EIA).

Reference Black Level. The picture signal level corresponding to the specified maximum limit for black peaks. (EIA).

Reference White Level. The picture signal level corresponding to the specified maximum limit for white peaks. (EIA).

Resolution. The details that can be distinguished on the television screen. Vertical resolution is a function of the number of scanning lines one sees on the screen. Horizontal resolution is a function of the number of intensity variations within each scanning line, and is generally variable according to the bandwidth of the system in use.

Resolution:

ANGULAR. The angle subtended by the image of a point on the object.

HORIZONTAL. The amount of resolvable detail in the horizontal direction in a picture. It is usually expressed as the number of distant vertical lines, alternately black and white, that can be seen in a distance equal to the picture height. This information usually is derived by observation of the vertical wedge of a test pattern. A picture that is sharp and clear and shows small details has good or high resolution. If the picture is soft and blurred and small details are indistinct, it has poor or low resolution. Horizontal resolution depends upon the high-frequency amplitude and phase response of the pickup equipment, the transmission medium, and the picture monitor as well as the size of the scanning spots. (EIA).

PHOTOGRAPHIC.  The number of lines per inch or per millimeter which can be resolved by an optical system.

VERTICAL.  The amount of resolvable detail in the vertical direction in a picture. It is usually expressed as the number of distinct horizontal lines, alternately black and white, that can be seen in a test pattern. The vertical resolution is fundamentally limited by the number of horizontal scanning lines per frame. Beyond this, vertical resolution depends on the size and shape of the scanning spots of the pickup equipment and picture monitor and does not depend upon the high-frequency response or bandwidth of the transmission medium or picture monitor. (EIA).

RETAINED IMAGE (IMAGE BURN).  A change produced in or on the target that remains for a large number of frames after the removal of a previously stationary light image, and which yields a spurious electrical signal corresponding to that light image. (EIA).

RINGING (IN RECEIVERS).  An oscillatory transient occurring in the output of a system as a result of a sudden change in input. (EIA).

ROLL-OFF.  A gradual decrease in signal voltage, usually associated with an increase in frequency.

SATURATION:

IN COLOR DISPLAYS.  The degree to which a color is pure, undiluted with white light.

IN AMPLIFIERS.  The point on the operational curve of an amplifier at which an increase in input amplitude will no longer result in an increase in amplitude at the output of an amplifier. An amplifier so overextended is said to be saturated.

SCANNING.  Moving the electron beam of a pickup tube or a picture tube diagonally across the target or screen area of the CRT.

SENSITIVITY (OF A CAMERA TUBE).  The signal current developed per unit of incident radiation density (i.e., per watt per unit area). Unless otherwise specified, the radiation is understood to be that of unfiltered incandescent source of 2870 degrees K, and its density, which is generally measured in watts per unit area, may then be expressed in foot-candles. (IRE).

SETUP. The ratio of the difference between black level and blanking level to the difference between reference white level and blanking level, usually expressed in per cent.

SHADING. A large area brightness gradient in the reproduced picture, not present in the original scene. (EIA).

SIGNAL-TO-NOISE RATIO. The ratio of the value of the signal to that of the noise. (EIA).

(CAMERA TUBES). The ratio of peak-to-peak signal output current to RMS noise in the output current. (EIA).

SPIKE. A transient of short duration, comprising part of a pulse, during which the amplitude considerably exceeds the average amplitude of the pulse.

SQUAREWAVE RESPONSE. The ratio of (1) the peak-to-peak signal amplitude given by a test pattern consisting of alternate black and white bars of equal widths to (2) the difference in signal between large-area blacks and large-area whites having the same illuminations as the black and white bars in the test pattern. (EIA).

STREAKING. A term used to describe a picture condition in which objects appear to be extended horizontally beyond their normal boundaries. This will be more apparent at vertical edges of objects when there is a large transition from black to white or white to black. (EIA).

SWITCHER-FADER. A control that permits each of two or more cameras to be selectively fed into the distribution system. The fader permits gradual transition from one camera to another.

SYNC. A contraction of synchronization, synchronizing, or synchronous.

SYNC COMPRESSION. The reduction in the amplitude of the sync signal, with respect to the picture signal, occurring between two points of a circuit. (EIA).

SYNCHRONIZED MOTION PICTURE PROJECTOR. A motion picture projector that is specially equipped with a speed and shutter mechanism compatible with the television frame and scanning system. The use of non-synchronous projectors ordinarily results in interference patterns on the reproduced images.

SYNCHRONIZING. Maintaining two or more scanning processes in phase. (EIA).

SYNCHRONOUS DEMODULATION. In a color television receiver, the process of separately detecting the *I* and *Q* sidebands of the color subcarrier system.

SYNC LEVEL. The level of the peaks of the sync signal. (EIA).

SYNC SIGNAL. The signal employed for the synchronizing of scanning. (EIA). Added to the composite video waveform, it is usually separated by circuitry in the monitor and used to synchronize the deflection generators.

TALLY LIGHT. Signal lights installed at the front and rear of television cameras to inform performers and crew members when a particular camera is on the air.

TARGET (CAMERA TUBES). A structure employing a storage surface which is scanned by an electron beam to generate a signal output current corresponding to a charge-density pattern stored thereon. (EIA).

TEARING. A term used to describe a picture condition in which groups of horizontal lines are displaced in an irregular manner. (EIA).

TEST PATTERN. A chart especially prepared for checking overall performance of a television system. It contains various combinations of lines and geometric shapes. The camera is focused on the chart, and the pattern is viewed at the monitor for fidelity. (EIA).

VERTICAL RETRACE. The return of the electron beam to the top of the picture tube screen or the pickup tube target at the completion of the field scan.

VIDEO. A term pertaining to the bandwidth and spectrum position of the signal resulting from television scanning. In current usage, video means a bandwidth in the order of megahertz, and a spectrum position that goes with a dc carrier. (EIA).

VIDEO AMPLIFIER. A wideband amplifier for the picture signals.

VIDEO SIGNAL (NONCOMPOSITE). The picture signal. A signal containing visual information and horizontal and vertical blanking. See "Composite Video Signal."

VIDICON. A camera tube in which a charge-density pattern is formed by photoconduction and stored on that surface of the photoconductor that is scanned by an electron beam, usually of low-velocity electrons. (EIA).

VIEWFINDER. Basically, a small television monitor that is built into the television camera and picks up only the picture from that particular camera, thus enabling the cameraman to see exactly what his camera is "seeing."

VTR. Video tape recorder.

WAVEFORM MONITOR. An oscilloscope designed especially for viewing the waveform of a video signal.

WHITE COMPRESSION (WHITE SATURATION). The reduction in gain applied to a picture signal at those levels corresponding to light areas in a picture with respect to the gain at that level corresponding to the midrange value in the picture. (EIA).

Y SIGNAL. A signal transmitted in color television containing brightness information. This signal produces a black and white picture on a standard monochrome receiver. In a color picture, it supplies fine detail and brightness information.

ZOOM. To enlarge or reduce on a continuously variable basis, the size of a televised image. It may be done electronically or optically.

ZOOM LENS. A special camera lens whose focal length or angle of view is continuously adjustable, while the lens remains always in focus on an object or scene within its wide range.

# QUESTIONS KEYED TO

# CHAPTERS 1 THROUGH 16

These questions may be utilized to check whether you have understood and retained the material within the respective chapters.

## CHAPTER ONE

**1.** What is probably the greatest difficulty encountered with television cameras used for out-of-doors surveillance?

**2.** What is the major requirement for a television camera used in a data transmission application?

**3.** Name three desirable characteristics of a television camera that is to be used in an area of high atomic radiation.

**4.** Why is stainless steel often used as a camera housing for underwater applications?

**5.** Why are wide-angle lenses normally used in underwater television cameras?

**6.** What difficulties arise when a television camera is used at high altitude?

**7.** What advantages are realized using "slow scan" cameras in spaceborne applications?

## CHAPTER TWO

**1.** When is it desirable to use a two-unit camera (separate camera head and camera control unit)?

**2.** What is the purpose of a sync generator?

**3.** What are the advantages of driving several camera scanning systems with a single sync generator?

**4.** How does a television monitor differ from a television receiver?

**5.** Why are bandwidths of closed circuit television systems not restricted as are those in broadcast applications?

**6.** What is the purpose of a distribution amplifier?

**7.** What is the purpose of a remote pan-tilt unit?

## CHAPTER THREE

**1.** Define *Index of Refraction*.

**2.** What is meant by the term *focal point?*

**3.** When an object is moved toward a lens, in which direction will the focal point tend to move? (See Figure 3-6.)

**4.** Define *focal length*.

**5.** In vidicon television cameras, which lens is considered to have the "normal" magnification or field of view?

**6.** What is the width of field of a 1-inch lens viewing a scene that is 100 feet distant?

**7.** What is the function of a lens iris?

**8.** Define *lens speed*.

**9.** If a lens is adjusted to an iris setting of F/8, and the iris opening is ⅛ inch, what is the lens focal length? (See Figure 3-11.)

**10.** What effect does lens iris setting have on the depth of field?

**11.** Are most lens abberations more pronounced with the iris in the wide open position, or at a reduced setting?

**12.** If a 5:1 zoom lens has an angle of view of 5 degrees at its maximum focal length, what will be the field of view at its minimum focal length?

## CHAPTER FOUR

**1.** Name the three major sections of an image orthicon camera tube.

**2.** What is the purpose of the multiplier section in an image-orthicon?

**3.** Is an image-orthicon more sensitive to red or blue light?

**4.** Why are the resistor/capacitor networks included at each DC input to the image-orthicon? (See Figure 4-6.)

**5.** What is the purpose of the *target* in a vidicon tube?

**6.** What electrical parameter of the vidicon is varied to change its effective sensitivity?

**7.** Describe *dark current* in a vidicon.

**8.** As the number of resolution elements in a scene being viewed by a vidicon increases, what is the effect upon the output signal of the tube?

**9.** Why is no gamma correction needed when the vidicon is used in monochrome cameras designed for general closed circuit usage?

**10.** The standard vidicon is most sensitive to what colors?

## CHAPTER FIVE

**1.** What is meant by camera tube *blanking?*

**2.** Why is the horizontal scanning frequency higher than the vertical scanning frequency?

**3.** While it is desirable to employ a low vertical scanning rate, it cannot be much below 60 hertz for most uses. Why?

**4.** Why was the 60-hertz vertical scanning frequency chosen as the standard in the United States?

**5.** Define the terms *interlace, field,* and *frame* as they apply to the television raster.

**6.** Define *vertical resolution.*

**7.** Define *horizontal resolution.*

**8.** What is the generally accepted Kell factor that is used in computing vertical resolution?

**9.** What is the vertical resolution capability of a television raster having 800 active horizontal scanning lines?

**10.** What basic reference is used to define horizontal or vertical resolution in television systems?

**11.** What two methods of deflection are used to achieve beam scanning in television camera tubes? Which is the more commonly used?

**12.** What type of current waveform is necessary to achieve a linear beam scanning excursion?

**13.** In the simple UJT oscillator shown in Figure 5-13, what would be the effect of an increase in the size of capacitor $C_t$?

**14** What is the purpose of diode $CR1$ in Figure 5-17?

**15.** What is the purpose of the *cable delay* circuitry in Figure 5-20?

**16.** Why is vidicon protection circuitry included in a great majority of television cameras?

## CHAPTER SIX

**1.** What is the purpose of the video preamplifier in a two-unit camera system?

**2.** Why is the first stage of the video preamplifier located as near the vidicon target as possible?

**3.** Why is it desirable to achieve the majority of system gain in the first few stages of amplification?

**4.** Describe the action of video peaking circuits in video amplifier and processing systems.

**5.** What is the purpose of *aperture correction?*

**6.** Name two common methods of achieving aperture correction.

**7.** Explain the effect of clamping circuits upon the video signal.

**8.** To what reference level is the video clamped in Figure 6-11?

**9.** What is the purpose of the blanking signal inserted onto the video waveform? How does this blanking signal differ from that utilized in the camera head?

**10.** How may a negative image be obtained on a television monitor?

**11.** How is gamma correction achieved in video amplifiers? Why is it needed?

**12.** Why is a variable gamma correction not incorporated in many monochrome vidicon camera systems?

**13.** In Figure 6-20, how does variation in emitter resistance vary gamma?

**14.** What is the purpose of $CR204$ in the auto-target circuit shown in Figure 6-21?

**15.** To use a closed circuit television camera in conjunction with a standard TV receiver, what circuitry must be incorporated in the camera to achieve compatibility?

**16.** What is the purpose of $C313$ in Figure 6-23?

**17.** When the arm of potentiometer $R324$ (Figure 6-23) is positioned to the collector of $Q305$, is aperture correction maximum or minimum?

## CHAPTER SEVEN

**1.** What is the purpose of a sync generator in a television system?

**2.** What is the purpose of using countdown circuitry to derive a vertical drive signal that is an exact submultiple of the horizontal frequency?

**3.** What is meant by *random interlace?*

**4.** Why is an emitter follower used as the output to the vertical yoke in Figure 7-1?

**5.** Describe the relationship between horizontal blanking, horizontal sync, front porch, and back porch in a composite video waveform.

**6.** Name the four types of pulses that occur during the vertical blanking interval of a broadcast-type video signal.

**7.** What is the purpose of the serration pulses?

**8.** Why do equalizing and serration pulses occur at twice the horizontal frequency?

**9.** Which EIA recommended video waveform is the standard for broadcast television in the U.S.?

**10.** What four output signals are normally provided by a sync generator that conforms to the recommendations of EIA standard RS-170?

**11.** Are equalizing and serration pulses a necessary part of composite video waveforms that comply with EIA standards RS-330 or RS-343?

**12.** What is the purpose of $Q1$ and $Q5$ in Figure 7-14?

**13.** Explain the function of $Q3$ in Figure 7-14.

**14.** What would be the effect upon circuit output if $Q104$ were defective in Figure 7-17?

**15.** In Figure 7-20, how would an increase in the size of $B_b$ affect the output pulse width of the boxcar circuit?

## CHAPTER EIGHT

**1.** How are television monitors generally classified?

**2.** How is the high voltage for the picture tube usually generated in monochrome television monitors?

**3.** How is the intensity of the spot brightness on a monochrome picture tube usually varied?

**4.** What must be the polarity of input pulses in Figure 8-5 to achieve proper trigger action?

**5.** Describe sawtooth development in Figure 8-7.

**6.** Explain the action of the *vertical hold* control in Figure 8-7.

**7.** What is the purpose of $R328$ and $C311$ in Figure 8-8?

**8.** In Figure 8-10, how is the AFC bias voltage obtained for the oscillator?

**9.** Why are two series transistors used in the example of Figure 8-12?

**10.** In Figure 8-13, why is such a small value of capacitance ($C7$) sufficient to filter the high voltage?

**11.** What is the function of a sync separator circuit?

**12.** What is normally used as a reference in the clamping circuits of most monitors? Why?

**13.** What is an advantage in employing a differential input to a monitor video amplifier?

## CHAPTER NINE

**1.** What three types of focus are necessary on most vidicon television cameras?

**2.** What is the normal amplitude of a composite video signal intended for broadcast use? For use in CCTV?

**3.** In what terms are the video and noise voltages expressed when measuring signal-to-noise?

**4.** What is a *weighting filter* as used when making signal-to-noise measurements?

**5.** What advantages might be realized by using a detector probe when sweeping a television system?

**6.** What is the purpose of a dot-bar generator in checking the linearity of a television camera?

**7.** What is the effect upon a monitor presentation when a vidicon is underscanned?

**8.** Why is it not a good practice to underscan a vidicon?

**9.** What is the purpose of an *equalizing* amplifier?

## CHAPTER TEN

**1.** What characteristic of visible light defines its color?

**2.** What is meant by the term *monochromatic* light?

**3.** Describe the action of a colored filter upon white light.

**4.** What color results from a combination of red and green light?

**5.** List the relative brightness of red, blue, and green needed to form white light.

**6.** When red, blue, and green objects of equal size are moved into the distance, which will appear to lose its color first?

**7.** Does the phosphor dot screen on modern television receivers utilize the principles of color by addition or color by subtraction?

**8.** What is the purpose of dichroic mirrors in a color camera?

**9.** What is the frequency of the color subcarrier used in the NTSC color system?

**10.** What is meant by *frequency interleaving* in the color video signal?

**11.** What is the horizontal scanning frequency for color systems in the United States? The vertical?

**12.** Why is the *G-Y* signal not used to develop the composite color signal?

**13.** In the NTSC color signal, why are *I* and *Q* signals used instead of *R-Y* and *B-Y* signals?

**14.** What is the purpose of the color burst signal?

## CHAPTER ELEVEN

**1.** What is the purpose of the additional camera tube in a four-tube color camera system?

**2.** Why must shading generators be used in color cameras?

**3.** Explain the purpose of gamma correction circuits in color cameras.

**4.** Explain the need for electronic signal enhancement in three-tube color cameras.

**5.** In the example of signal enhancing circuitry illustrated in Figure 11-9, why is the video modulated onto an RF carrier prior to the delay networks?

**6.** Why is the *detail* signal generated by the signal enhancing circuits added to the green channel instead of the red or blue channel?

## CHAPTER TWELVE

**1.** How is a constant phase relationship established between the color subcarrier and the horizontal and vertical frequency and phase relationships?

**2.** What is the purpose of *sync lock* or *genlock* circuits?

**3.** What is meant by the term *breezeway* as it applies to the color video signal?

**4.** Why are the output pulses from multivibrating $Q21$, $Q22$ (Figure 12-2) designed to be three-fourths the width of a horizontal line?

**5.** What is the purpose of $Q17$ in Figure 12-4?

**6.** What determines pulse amplitude output from the circuit in Figure 12-6?

**7.** What is the purpose of the *burst flag* pulse?

**8.** Approximately how many cycles of subcarrier frequency are contained in the color burst?

## CHAPTER THIRTEEN

**1.** What is the purpose of a color encoder or modulator?

**2.** Why are delay circuits $DL1$ and $DL2$ included in the circuitry of Figure 13-2?

**3.** By what process are the $RGB$ signals converted into the $I$, $Q$, and $Y$ signals?

**4.** What is the purpose of the red and green inverters in Figure 13-3?

**5.** When the $I$ input to the circuit of Figure 13-5 is negative, what is the operating condition of the diode bridge $CR6$ through $CR9$?

**6.** The polarity of the sync signal applied to the video outputs 1 and 2 (in Figure 13-8) is negative going. What then is the polarity of the sync signal at the base of $Q20$?

**7.** What is the purpose of a color-bar generator?

**8.** In a color-bar generator, what must be the output configuration of the $R$, $G$, and $B$ signals to realize the color Cyan on the monitor presentation?

## CHAPTER FOURTEEN

**1.** What is the difference between an *RGB* color monitor and a *decoding* color monitor?

**2.** What is the purpose of the *mixer* in a television receiver tuner section?

**3.** What is the purpose of AGC in television receivers?

**4.** What determines the frequency of the RF oscillator in a television receiver?

**5.** Why must the sound subcarrier be suppressed prior to the video detector stages?

**6.** What is meant by *keyed* AGC?

**7.** Explain the principles of noise cancellation in video receivers.

**8.** List the composition of the $I$ and $Q$ signals in terms of $R$-$Y$ and $B$-$Y$.

**9.** Under what two conditions will the output of the demodulator circuit of Figure 14-14 be zero?

**10.** Express $G$-$Y$ in terms of $R$-$Y$ and $B$-$Y$.

**11.** Why is the $Y$ signal added to the $R$-$Y$, $B$-$Y$, and $G$-$Y$ signals in the color signal?

**12.** What is the purpose of *color killer* circuitry in a color receiver?

**13.** What is meant by *convergence* in a color picture tube?

**14.** In general, what two types of convergence procedures must be performed on a color television set?

## CHAPTER FIFTEEN

**1.** What is the difference between a *gap* and a *lap* type of video switching transition?

**2.** In terms of inputs and outputs, what is meant by a $5 \times 10$ switcher?

**3.** Explain the general differences between distribution-type and broadcast-type video switchers.

**4.** What is meant by *vertical-interval* switching?

**5.** What is the purpose of an *effects* generator?

**6.** Describe a *single re-entry* and a *double re-entry* switching system.

**7.** What is the purpose of a *sync adder* in a switching system?

**8.** What signals are usually used as clock pulses for vertical interval switching?

## CHAPTER SIXTEEN

**1.** Why are rotating scanning heads usually used in video tape recording?

**2.** List three types of tape head scanning commonly used in video tape recorders.

**3.** Why is the video signal frequency modulated prior to recording on tape?

# SUBJECT INDEX

C

## F

*M*

*Q*

*R*